The Human Evolutionary Transition

The Human Evolutionary Transition

FROM ANIMAL INTELLIGENCE TO CULTURE

MAGNUS ENQUIST,

STEFANO GHIRLANDA,

AND JOHAN LIND

PRINCETON UNIVERSITY PRESS
PRINCETON & OXFORD

Published by Princeton University Press
41 William Street, Princeton, New Jersey 08540
99 Banbury Road, Oxford OX2 6JX

press.princeton.edu

ISBN 978-0-691-24075-6
ISBN (pbk.) 978-0-691-24077-0
ISBN (e-book) 978-0-691-24076-3

British Library Cataloging-in-Publication Data is available

Editorial: Alison Kalett and Hallie Schaeffer
Production Editorial: Jill Harris
Cover Design: Heather Hansen
Production: Lauren Reese
Publicity: Matthew Taylor and Charlotte Coyne
Copyeditor: Jennifer McClain

Cover image © Anita Ponne / Shutterstock

This book has been composed in Arno and Sans

10 9 8 7 6 5 4 3 2 1

CONTENTS

ACKNOWLEDGMENTS

We worked together on this book for about 15 years, during which time we have benefited from countless inspiring conversations with our colleagues at the Centre for the Study of Cultural Evolution (CEK) at Stockholm University, Sweden, at the Department of Psychology at Brooklyn College, and at other institutions. We are especially grateful to Alberto Acerbi, Elliot Aguilar, Hanna Aronsson, Staffan Bergwik, Andrew Buskell, Matthew Crump, Andrew Delamater, Ida Envall, Kimmo Eriksson, Guillermo Esber, Daniela Fuchs, Alexander Funke, Frank Grasso, Gustaf Gredebäck, Anandi Hattiangadi, Jerry Hogan, Fredrik Jansson, Arne Jarrick, Anna Jon-And, Markus Jonsson (the main architect of our learning simulator; see section 1.5.5), Martin Kolk, Kerstin Lidén, Patrik Lindenfors, Eva Lindström, Matts Lindström, Jérôme Michaud, Hanna Müller, Laila Nauman, Jared Taglialatela, Hans Temrin, Vera Vinken, Maria Wallenberg-Bondesson, and Andreas Wartel.

We also thank four anonymous reviewers, for suggesting many improvements and clarifications; Leora Fox, for carefully and constructively editing our first draft; and the Departments of Zoology and Archaeology and Classical Studies at Stockholm University, for their long-term support.

We gratefully acknowledge financial support from the Knut and Alice Wallenberg Foundation (grant 2015.0005), MISTRA (grant 1228811), and Stockholm University. Stefano has also been supported by a fellowship leave from Brooklyn College and a CUNY Graduate Center Midcareer Fellowship.

This book would not have been possible without the help and encouragement of our families, to whom it is dedicated with love.

The Human Evolutionary Transition

1

Challenges to the Evolution of Intelligence

- The world offers many resources to organisms, but it is also large and complex.
- Obtaining resources is hard. Exploration takes time, behavior may have delayed consequences, and informative stimuli are often mixed with noninformative ones.
- Animals have evolved several solutions to these challenges, such as learning by trial and error and learning from others.
- In humans, a new solution has emerged that couples cultural information and the capacity to think.
- The transition from animal to human intelligence can be understood by reasoning about *sequences*: sequences of behavior, sequences of information-processing steps, and sequences of stimuli.

1.1 What Happened in Human Evolution?

Until three or four million years ago, our ancestors inhabited a small region in Africa and were probably similar in intelligence to contemporary great apes. Since then, our species has acquired many unique features and has colonized almost all terrestrial habitats. Figure 1.1 shows a coarse summary of our species' history. Characteristics that emerged in human evolution include language, complex societies, material and nonmaterial culture, such as art and science, and a rich inner world of thoughts, hopes, and fears. While it is clear that our species' ability to process and organize information has changed, it is not easy to pinpoint what the changes were and what caused them. We refer to these changes collectively as the *human evolutionary transition*, in analogy with other momentuous evolutionary events, such as multicellularity and sexual reproduction (Maynard Smith and Szathmáry 1995). In this book, we put forward a theory of the *content* of the transition (how human

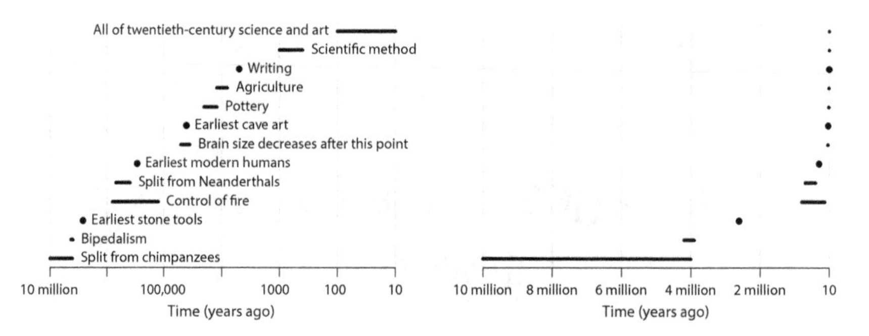

FIGURE 1.1. Some major developments in human history. The two panels include the same events. On the left, a logarithmic time axis enables us to label events clearly. On the right, a linear axis highlights the dramatic speed of cultural evolution: all of twentieth-century science and art has been developed during a mere 0.001% of the time depicted, or about 0.1% of the time since the appearance of modern humans. Data sources: split from chimpanzees: 10–4 million BP (years before present; Dolhinow and Sarich 1971, White et al. 2009); bipedalism: at least since *Australopithecus afarensis*, around 4 million BP ("Lucy"; Ward 2002); earliest stone tools: 3.3 million BP (Harmand et al. 2015); control of fire: between 800k (Berna et al. 2012) and 125k BP (Karkanas et al. 2007); split from Neanderthals: around 600k BP (Schlebusch et al. 2017); earliest modern humans: around 300k BP, based on population genetics (Schlebusch et al. 2017) and fossils (Hublin et al. 2017); brain size decrease: 50–30k BP (Henneberg 1988); cave art: at least 40k BP (Aubert et al. 2014, Brumm et al. 2021); pottery: at least 20k BP (Wu et al. 2012); agriculture: around 10k BP (Vasey 2002); writing: at lea1st 5k1 BP (cuneiform: Walker 1987); scientific method: conventionally, from eleventh-century Arab scholars to seventeenth-century Europeans.

information processing differs from that of other animals) and of its *causes* (the evolutionary events that caused the transition, including genetic and cultural evolution).

In this chapter, as an introduction to a broader argument about the nature of animal and human intelligence, we consider general challenges to the evolution of intelligence. Before beginning, we note that our use of the word *animals* typically excludes humans. Humans are animals, but this usage is convenient to compare humans and other animals.

1.2 The World Is Full of Opportunities

The world is incredibly rich in resources that organisms can exploit, where "resource" is broadly intended as anything that can aid survival and reproduction. Tropical habitats are spectacularly replete with life, but even in hostile environments, such as deserts or the Arctic, organisms can extract enough energy from their surroundings to sustain their species. The behaviors that animals use to obtain resources vary greatly in complexity, depending on what resources they rely on. Filter feeders, for example, rely on resources that are continuously available and floating in water, and therefore their feeding behavior is limited to forcing water through specialized structures that capture nutrients. This simple strategy has evolved many times, such as in molluscs, crabs, shrimp, sharks, and whales. Most animals, however, rely on more complex strategies. Predators may use stealth, speed, venom, or an artifact such as a spider web. Animals that rely on seasonal resources must secure an energy store for hard times, by accumulating fat, hoarding food, or decreasing energy consumption. These are just a few examples of the astonishing diversity of animal life.

Exploiting environmental resources requires organized sequences of actions, in all but the simplest cases. This concept is crucial to the thesis of our book. Weaving a web, hunting prey, or escaping a predator requires executing a sequence of behavior, often with great precision. A mistake results frequently in a lost opportunity, or even in serious harm. Animals have discovered a staggering number of productive sequences of behavior, among many potential alternatives that do not work or that work less well. Humans, however, have developed productive sequences of incredible length. To better appreciate the gap between human and nonhuman sequences, consider a sophisticated nonhuman tool, such as a twig used by chimpanzees to extract termites from their nest, and an outwardly unsophisticated human tool, such as a coat hanger. We are fascinated by chimpanzees' abilities for tool manufacture, but we are typically indifferent to coat hangers. Chimpanzee tools, however, can be built by a single individual with a few actions, such as locating an appropriate twig, detaching it from the branch, and stripping it of its leaves (Nishihara et al. 1995, Sanz et al. 2009). In contrast, the sequence of actions that goes into building a coat hanger is so long as to be untraceable. The coat hanger is made by machines with thousands of parts (figure 1.2), using metal that has ultimately been obtained through an organized mining operation involving thousands of people. The same holds for all but perhaps the simplest objects of daily life in industrial societies (Jordan 2014). To understand why only humans have been able to discover such long sequences, we must first examine why finding productive sequences is difficult at all.

1.3 Sequences and Combinatorial Dilemmas

The design of a behavior system determines what resources it can exploit. For example, an organism whose behavior is completely hardwired is unable to adjust to new food sources or new threats. Likewise, an organism with poor senses might fail to distinguish edible from inedible food. Accordingly,

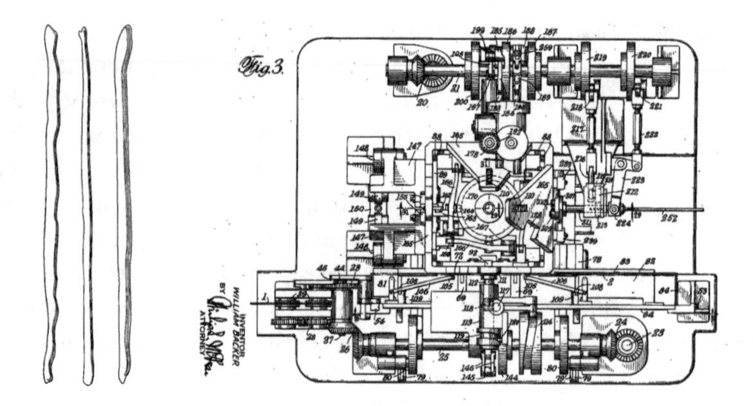

FIGURE 1.2. Left: Twigs used by chimpanzees to perforate termite mounds (redrawn based on Nishihara et al. 1995). Right: One of 32 diagrams describing a machine to manufacture coat hangers (US patent 2,041,805, November 23, 1935).

behavior systems face many challenges. Sense organs must be good enough to perceive relevant information, memory good enough to store such information, and information processing sophisticated enough to drive efficient behavior while also being sufficiently fast. The challenges that are most important for this book are several kinds of combinatorial dilemmas involving sequences. These are difficulties that derive from the exponential increase in possibilities when a task grows in complexity (Bellman 1961, Dall and Cuthill 1997, Keogh and Mueen 2010). In our case, the dilemmas arise when going from shorter sequences to longer sequences. The three combinatorial dilemmas that play a fundamental role in this book involve sequences of behavior, sequences of information-processing steps, and sequences of stimuli.

1.3.1 Behavioral Sequences

In principle, it is possible to find productive sequences of behavior through brute-force exploration, that is, by simply trying out all possible sequences. This, however, is too time consuming to be generally practical. It's like trying to phone someone by dialing all possible phone numbers. More formally, consider an animal exploring sequences of l actions, and assume that each action is chosen randomly out of a repertoire of m. There are then m^l sequences that the animal can try out. If r of these sequences are rewarding, the expected number of attempts before finding a rewarding sequence is

$$\frac{m^l}{r} \tag{1.1}$$

This number increases for longer sequences and more actions. Finding longer sequences is *exponentially* harder than finding shorter ones. Finding a sequence that is just 3 actions long, with a behavioral repertoire of 10 actions, would take $10^3 = 1000$ attempts. Finding a sequence of 10 actions would take $10^{10} = 10$ billion attempts. Furthermore, time spent learning detracts from time spent using what has been learned (the exploration-exploitation dilemma, considered in sections 3.2 and 4.3).

Equation (1.1) also highlights a difficulty with increasing motor flexibility. Most animals have a rather constrained repertoire of possible actions, which is strongly determined genetically (Tinbergen 1951, Eibl Eibesfeldt 1975). Humans and other primates are remarkable for their flexibility, and humans in particular for their ability to learn a diversity of complex skills, such as gymnastics, surgery, or playing the piano. Motor flexibility appears advantageous because it increases the animal's capacity to act on its environment, but it also leads to learning costs that compound with sequence length. For example, an animal that can perform 10 actions can try out $10^2 = 100$ sequences of 2 actions, but this number increases to $50^2 = 2500$ for an animal that can perform 50 actions, yielding a 25-fold increase in the time to try out all possible actions. Thus motor flexibility is not necessarily advantageous, and we can add it to the list of human features whose evolution we would like to understand.

1.3.2 *Mental Sequences*

Most animals need fast decision making. A foraging bird, for example, is constantly deciding where to hop next, whether to look up to check for predators, whether to switch to a different activity, whether to attack a bug it has spotted, and so on. A more complex information-processing mechanism can be useful if it makes better decisions, but not if this takes too much time. To see how combinatorial dilemmas can arise in information processing, consider planning a sequence of actions. If there are $m = 10$ available actions, planning a single action requires considering 10 alternatives, but planning an action sequence incurs exponentially increasing costs, similar to brute-force exploration. Planning 2 actions may require considering up to $m^2 = 100$ alternatives, and planning 3 actions up to $m^3 = 1000$. Even if imagining actions can be faster (and less risky) than actually performing them, planning sequences can still be prohibitively time consuming. We consider the costs of planning in chapter 11.

1.3.3 *Stimulus Sequences*

To behave efficiently, organisms typically require information from the environment. Locating food, finding mates, and avoiding predators are some of the activities in which environmental information is essential. Extracting and using information from the stream of sensory experiences poses many challenges. One is stimulus recognition. For example, an organism can only learn to eat a particular fruit if it can recognize sensory states that indicate the presence of that fruit. These sensory states are numerous, arising from differences between individual pieces of fruit as well as differences in background, viewpoint, distance, and lighting conditions. This problem is typically solved through stimulus generalization, that is, by evaluating the current stimulus based on its similarity to familiar stimuli (stimuli to which a response has already been established). The rules of stimulus generalization appear similar regardless of whether knowledge of the familiar stimuli is inborn or learned (Ghirlanda and Enquist 2003).

Stimulus recognition and generalization are very difficult computational problems, as witnessed by the fact that computer systems are just now becoming competent at recognizing objects in the real world (Klette 2014). However, we believe that there are no major differences between humans and animals in this domain, and for this reason we do not devote much space to the topic, with two important exceptions. The first is the recognition of *sequences* of stimuli, for which animals' abilities do appear more limited than humans' (chapter 5). The second is the possibility of learning new *representations* of stimuli, such as symbolic representations or representations that emphasize meaningful features, which we also think is much more limited in animals than in humans (chapter 13). We believe that these differences between animals and humans are rooted in the following combinatorial dilemma.

Organisms experience a continuous stream of stimuli, and have to strike a balance between the potential advantages of remembering more information to use for learning and decision making and the costs that stem from remembering more information. One cost is that more information requires a larger memory. The main cost, however, may be that remembering more might actually make learning and decision making more difficult, rather than easier. To appreciate this fact, suppose that an animal can perceive s different stimuli, and that it decides what to do based on the last n stimuli perceived. If $n = 1$, the animal uses the current stimulus only. It thus needs to know what to do in s different situations. If $n = 2$, the animal remembers the current stimulus and the previous one. This may reveal more information about the environment, but it also means that there are now s^2 situations (s^2 pairs of stimuli) in which the animal must know what to do. With increasing n, the number of potential situations increases exponentially as s^n. Even if not all stimulus sequences actually occur, the animal is still faced with an increased number of possibilities that require a decision. With a realistic number of stimuli, the problem is daunting, and the only solution is to focus on a subset of the incoming

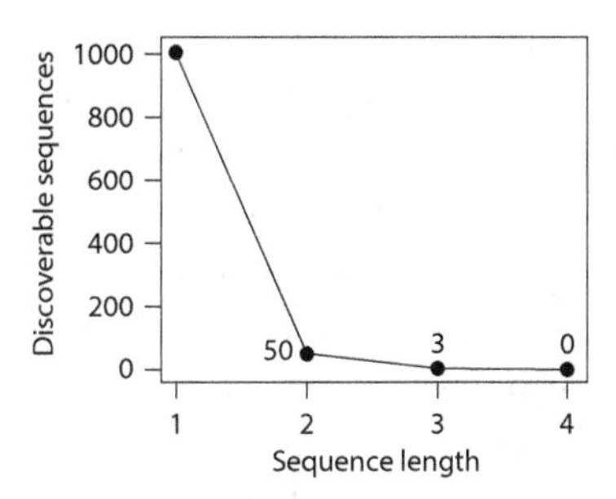

FIGURE 1.3. The number of productive sequences that can be discovered within a given time by brute-force exploration, as a function of sequence length. We assume that each action takes one time unit, so that discovering a sequence of length l takes on average lm^l time units, where m is the number of actions that can be performed. Thus the number of sequences that can be discovered in T time units is $n = \lfloor T/(lm^l) \rfloor$, with $\lfloor x \rfloor$ indicating the integer part of x. In the figure, $T = 10,000$ and $m = 10$.

information. The dilemma facing the animal is then what to remember and what to discard. We cover this important topic in chapter 5, where we argue that the dilemma lies behind many limitations in animal memory. Additionally, chapter 13 discusses how humans mitigate the problem by learning useful stimulus representations that are transmitted culturally.

1.4 How Can Combinatorial Dilemmas Be Managed?

Given the combinatorial dilemmas just highlighted, finding productive sequences of behavior appears prohibitively hard. With countless possibilities for experienced stimulus sequences and potential behavioral and mental sequences, animals and humans could accomplish very little without sound ways of managing the dilemmas (Zador 2019). Consider, for example, an animal that learns entirely by brute-force exploration, without any genetic or social information to help in the search for productive behavior. Suppose that the animal can perform $m = 10$ different actions and that it has time to try out $T = 10,000$ actions. Such an animal would have time to learn the best action in 1000 stimulus situations; but it could learn the best 2-action sequence in only 50 situations and the best 3-action sequence in only 3 situations, with no guarantee of learning the best 4-action sequence in even a single situation (figure 1.3). These discouraging numbers, moreover, apply under idealized conditions; for example, that the animal learns each correct sequence the first time it performs it.

We think that the evolution of both animal and human intelligence can be seen profitably as the discovery of strategies to manage combinatorial dilemmas (Zador 2019, Quiroga 2020). Animals and humans *can* survive and reproduce, hence they must have found viable strategies. For example, a generalist strategy might focus on learning many different short sequences, while a specialist strategy could focus on a few long ones whose learning could be supported by genetic predispositions. Only humans, it seems, have found a way to discover very many, very long sequences. This difference between humans and animals is the underlying theme of the book. As we will start to see in detail in chapter 2, our conclusion is that human and animal intelligence rely on two very different strategies for managing combinatorial dilemmas.

1.5 Our Approach

In this book, we aim to put forth a strong theory about the differences between human and animal intelligence. By "strong," we mean a theory that is formalized mathematically, from which clear empirical predictions can be derived. We believe that much of the disagreement and fragmentation that exist today in the fields of learning and cognition stem from a lack of formal theory. Human and animal behavior is often described with words that are inherently slippery: intelligence, cognition, understanding, planning, reasoning, insight, mental time travel, theory of mind, and others. These words describe abilities, but not how these abilities are achieved. In this book, we ask what organisms can do, what information their behavior is based upon, and how such information is acquired and used. This approach is not new. Formal models of human and animal behavior have been advanced in psychology, biology, and computer science. As detailed in chapter 2, we are greatly indebted to these traditions, as well as to the deep empirical and conceptual knowledge of behavior accumulated by ethologists and psychologists.

We do not expect all of our ideas to be correct, but we are convinced that our approach can refocus current debate in a direction with a more concrete promise of progress. The rest of this section elaborates our methodology.

1.5.1 What Is in a Mental Mechanism?

In this book, we try to understand how animals and humans arrive at productive behavior by focusing on the information-processing mechanisms that underlie behavior. We call any such mechanism *mental mechanism*. The term *mental* refers simply to the fact that these mechanisms operate within the brain, possibly disjointed from ongoing stimulation, and we do not attach to it any particular significance. We could have used equally well the terms

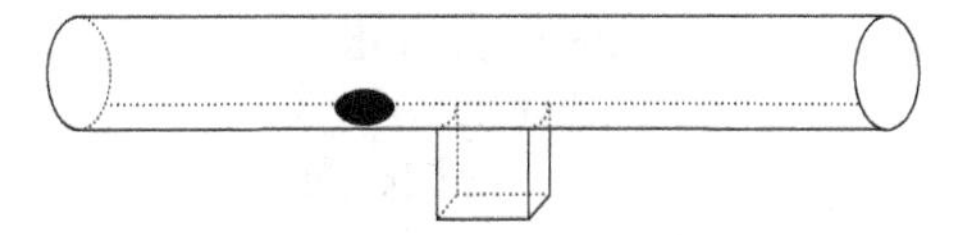

FIGURE 1.4. A trap tube. A food item (black oval) can be pushed out by inserting a stick from one end of the tube. Pushing from the other end causes the food to fall into the trap.

behavior mechanism (common in ethology) or *cognitive mechanism* (common in psychology), but the first may suggest a disregard for internal information processing, which is not our aim, and the second often excludes associative learning, which instead we include as a mental mechanism (see chapter 2).

We base our arguments on explicit models of mental mechanisms, by which we mean formal descriptions of how information is acquired, represented, updated, and used to make behavioral decisions. It is also important to consider which aspects of mental mechanisms develop rigidly (based primarily on genetic information) and which are instead flexible, that is, open to influence from experience. Ideally, all mechanisms should be mathematically defined in enough detail that they can be simulated on computers and implemented in robots. The advantage of this approach is to leave no doubt about how a mechanism is supposed to work, so that it is possible to determine what it can and cannot do.

We focus on behavior without addressing its neural basis. Neuroscience has made tremendous progress in relating behavior to nervous system operation, yet we are still far from understanding how differences in neural processes across species translate into different behavioral abilities. In other words, we start from behavior because we know what animals do much better than what their brains do. At the same time, the models we present in the book can be formulated readily as neural network models (Arbib 2003, Enquist and Ghirlanda 2005, Enquist et al. 2016), which may go some way toward closing the gap between brain and behavior.

1.5.2 *What Does It Take to Solve a Problem?*

Most conclusions about animal intelligence rest on experiments that challenge animals with various tasks or puzzles, and on inferring underlying mental mechanisms from the results of these experiments. Typically, these tasks can be described and analyzed in different ways. For example, a trap tube experiment (figure 1.4; see Visalberghi and Limongelli 1994, Limongelli et al. 1995) can be described as probing physical cognition, causal inference, or reasoning, or it can be seen simply as a choice between two actions (insert a stick from either end) based on visual information (where the food item lies). What matters, however, is what hypotheses about information processing are required to reproduce how animals behave on the task. To ascertain this conclusively, we see no alternative to formalizing the task and testing different mental mechanisms on it, comparing the behavior of models with that of animals. This often produces surprising results. For example, a task may turn out to be uninformative because it can be solved realistically by many mechanisms (trap tube experiments are one example). Of note, we find that associative learning—often deemed insufficient to reproduce certain aspects of animal intelligence—can behave realistically when modeled formally (see chapter 7).

1.5.3 What Are the Costs and Benefits of Mechanisms?

All mechanisms have costs in terms of time and energy needed to build and operate them, and in terms of the time they require to find and execute productive behavior. A mechanism that is more "intelligent" (can solve more problems) may actually be selected against if it needs too much energy, if it requires information that cannot be obtained easily, or if it takes too long a time to operate. We find that cost-benefit analysis of mechanisms is rare in current debate about animal cognition, yet such analysis is necessary to understand whether a mechanism is a viable evolutionary solution to a problem or set of problems. For this reason, we endeavor to understand the costs and requirements of the mechanisms we consider.

1.5.4 What Data Are Relevant?

Some readers may feel we have omitted studies or experimental paradigms that they deem important, while perhaps focusing on findings that are less central to current debate in animal cognition. Our selection of studies has been guided by our focus on the learning of productive behavioral sequences. For example, we have chosen not to discuss experiments on self-recognition in mirrors. These experiments may be important for a number of questions, but currently it is unclear what mechanisms are involved in self-recognition and what evolutionary advantages they bring.

We also focus on data that can be interpreted with as little ambiguity as possible. It is difficult, for example, to interpret the results of experiments whose subjects have an unknown developmental history, because a given behavior can often result from different mental mechanisms, given appropriate training. For this reason, observing a behavior without knowing the animal's experiences is often uninformative. Therefore, we favor experiments under controlled conditions and avoid, for the most part, observational studies or case studies of single individuals. One should also be aware that, even if we label one part of the experiment as "training" and another as "testing," animals learn from each and every experience. The possibility that behavior is influenced by learning during what we consider a mere test should always be taken into account. We see in chapter 7 that this consideration can be crucial in forming conclusions about mental mechanisms.

1.5.5 Learning Simulator and Online Script Repository

To achieve our goal of rigorous arguments, we have developed a simulation environment that we use throughout the book to specify experimental designs and mental mechanisms. The simulator is described in Jonsson et al. (2021) and is available at https://www.learningsimulator.org. It offers a programming language to specify environments in which learning agents can act in pursuit of goals, such as securing resources and avoiding danger, as well as a library of mental mechanisms that can be tested in any environment. Our simulation scripts can be found at https://doi.org/10.17045/sthlmuni.17082146.

1.6 Animal Rights and Human Responsibility

In this book, we conclude that nonhuman animals are less similar, cognitively, to humans than is often claimed (especially in nonacademic publications). We fear that this claim may be interpreted as detrimental to the humane treatment of animals, but this is not our intent. Rather, we stress that humans are the only species with the cognitive and technical capacity to willingly influence the fate of other organisms. With this power comes a responsibility for stewardship, which, to be effective, requires understanding the needs and capacities of animals (Stamp Dawkins 2008). Indeed, we suggest that species that are often considered less cognitively competent may actually be on the same footing as "more advanced" species considered worthy of ethical concerns. We discuss information-processing capacities only, rather than sentience or capacity to suffer, but these are not irrelevant to ethical arguments. For example, we conclude that, based on available evidence, all vertebrates may be able to assign positive or negative values to experiences (Macphail and Barlow 1985). What to make of this information is a complicated question subject to both ethical and practical considerations (Herzog 2010).

2

Seven Hypotheses

- We summarize the book in seven hypotheses on animal and human intelligence.
- Recent advances in machine learning have led to a new understanding of animal intelligence in terms of associative learning, sometimes at odds with common views.
- Animal intelligence consists of a sophisticated interplay between genetic predispositions and learning, which is superior to human intelligence in many situations.
- Human intelligence depends on internal reorganization of information (thinking) and needs culture to succeed.
- The human evolutionary transition is based on genetic changes that improved sequential information processing and increased mental flexibility.
- Cultural evolution has played a greater role in human evolution than often granted, including creating many uniquely human mental and behavioral skills.

2.1 Introduction

In this chapter, we present seven hypotheses about human and animal intelligence that we explore in the book.

Hypothesis 1: Humans and animals embody two different paths to intelligence, that is, two different strategies to find productive behavioral sequences amid combinatorial dilemmas. The differences are such that we regard the appearance of human intelligence as an evolutionary transition of similar magnitude to the evolution of multicellularity and sexual reproduction.

Like the hypotheses presented below, hypothesis 1 is both a synthesis of existing ideas and a controversial stance. To explain why we consider human intelligence an evolutionary transition, we have divided the book into two parts. In chapters 3–8, we detail our view of animal intelligence, summarized in section 2.2. Starting with chapter 9, we turn to discussing human intelligence, how it may have evolved, and why it evolved in just one species. These ideas are summarized in section 2.4.

2.2 Animal Intelligence

Animals are beautifully (if not perfectly) adapted to their environment, and intelligent behavior is fundamental to adaptation. In the first half of the twentieth century, ethology and behaviorist psychology made tremendous inroads into understanding animal intelligence, the first focusing mainly on innate aspects of behavior and the second on learned ones. Ultimately, however, they stopped short of producing a satisfactory theory. The cognitive revolution in psychology ushered in a new focus on information processing, at the same time that ethological insights into the organization of behavior were progressively integrated into psychological thinking. The most successful theory of intelligent behavior coming out of animal psychology is associative learning theory, which we cover extensively starting in chapter 3. However, there has always been skepticism about associative learning being able to explain all aspects of animal intelligence, leading many researchers to ascribe to animals abilities that we usually recognize in humans, such as planning, reasoning, and mental arithmetic. This stance was historically associated with Gestalt psychology and Tolman's purposive behaviorism, and is today apparent in the field of animal cognition (see section 2.2.2). It creates an intriguing paradox, in which animals have a cognitive sophistication similar to humans, while lacking the most characteristically human traits, such as complex language and culture.

We believe that recent developments in artificial intelligence may offer a resolution to this paradox, by explaining how animals can exhibit highly intelligent behavior without possessing the same mental abilities as humans. We explore two hypotheses about animal intelligence:

Hypothesis 2: Animal intelligence is generally based on a sophisticated interplay between a powerful form of associative learning, equivalent to some "reinforcement learning" algorithms from artificial intelligence, and genetic predispositions that guide both learning and decision making toward adaptive outcomes. These predispositions enable animals to learn without succumbing to combinatorial dilemmas.

Hypothesis 3: In many situations, the kind of intelligence described in hypothesis 2 is superior to human intelligence.

In this section, we outline key ideas from ethology, psychology, animal cognition, and machine learning and how these fields have contributed to our own ideas.

2.2.1 Ethology and Animal Psychology

Ethology and animal psychology have both studied animal intelligence and the underlying mental mechanisms. Ethology emerged out of biology and natural history, grounded in the theory of evolution (Tinbergen 1963). Its central idea is that behavior is organized into instincts that are biological adaptations to the environment (Tinbergen 1951, Hinde 1970). Thus ethology focused mainly on inborn aspects of behavior, such as inborn stimulus-response links and motivations like hunger and sex, and on how these serve survival and reproduction. The realization that "instinctual" behavior often depends on learning led to a partial revision of these ideas (Hogan 2001), yet inborn factors and biological adaptation remain central in ethology and allied fields, such as behavioral ecology and biological game theory (Krebs and Davies 1978, Maynard Smith 1982).

Animal psychology, on the other hand, is rooted in physiology and human psychology, and has typically studied mechanisms of learning, memory, and perception (Washburn 1908, Mackintosh 1983a, Bouton 2016, Pierce and Cheney 2018). Compared to ethologists, animal psychologists are traditionally less concerned with differences between species, with behavioral evolution, and with the biological function of behavior. Rather, their goal is to understand general laws of behavior that would hold across species, such as laws of perception, associative learning, and choice.

The ethological and psychological approaches to behavior have often clashed. For example, psychologists have criticized the concept of instinct, pointing to the importance of learning for the normal development of behavior (Lehrman 1953), while ethologists have replied that learning is genetically tailored to a species' natural environment (Hinde and Stevenson-Hinde 1973, Eibl Eibesfeldt 1975, Roper 1983). At the same time, many authors attempted a synthesis (Hinde 1970, Fantino and Logan 1979, Timberlake 1994, Hogan 2017), and most contemporary textbooks in psychology also refer to ethological work (Pearce 2008, Shettleworth 2010b, Bouton 2016).

One of the most influential research programs in animal psychology was behaviorism. Behaviorism sought to understand how behavior is affected by external variables, such as reinforcement history, without reference to the animal's internal workings (Skinner 1938, 1950). Behaviorism discovered many empirical laws of behavior and effective animal training methods, but it did not answer all the interesting questions about behavior, such as why a turtle and a blackbird respond differently to the same stimuli and questions about mental mechanisms. Behaviorists appreciated these questions (Skinner 1950, Skinner 1984) but left them unanswered. Ultimately, behaviorism's refusal to address internal mechanisms led to a renewed focus on "cognitive" explanations in psychology (Miller 2003, Watrin and Darwich 2012), while its perceived neglect of inborn factors made it unpopular among ethologists.

2.2.2 Animal Cognition and Associative Learning

Starting in the 1980s, research on animal learning and intelligence has increasingly been described as "animal cognition" research (Premack 1983, Roitblat et al. 1984, Dukas 1998, Balda et al. 1998, Bekoff et al. 2002, Shettleworth 2010b). The term *cognition* signals a focus on information processing that we also embrace. Considering how animals acquire, store, and use information is essential to understanding behavior and its evolution, and much successful work is based on this premise (Hassenstein 1971, McFarland 1971, Stephens and Krebs 1986, Dusenbery 1992). It may surprise some readers that we include associative learning theory among cognitive theories, even though it is commonly associated with reflex theory and behaviorism, which cognitive psychology rejected as limited and simplistic.

The origins of this muddle are manifold. Some early associative theorists did treat animals as simple stimulus-response machines (Thorndike 1911, Guthrie 1935), a position that leaves little room for intelligent behavior and that is still prominently covered in textbooks. Furthermore, the field of associative learning has continued to use the experimental methods and terminology of those early days, such as referring to learning as "conditioning." Ethologists' skepticism of learning psychology may also have contributed to rejecting associative models of animal intelligence. Yet modern associative learning theory is very much cognitive. Its main goal is to discover how animals represent and understand the world (Dickinson 1980, Rescorla 1988, Pearce 2008, Shettleworth 2010b, Haselgrove 2016). For example, the landmark model by Rescorla and Wagner (1972) uses environmental information to update internal variables (associative strengths) that capture correlations between events. The core computational principle of the model is often described in cognitive terms as involving "surprise," "prediction error," or "violation of expectation" (see chapter 3).

The classification of associative learning as "noncognitive" may have contributed to leading the field of animal cognition in a different direction. Namely, animals are attributed abilities that overlap considerably with those of humans, such as planning, reasoning, mental time travel, mental arithmetic, probability estimation, and others (Davis and Pérusse 1988, Griffiths et al.

1999, Bekoff et al. 2002, Blaisdell et al. 2006, Mulcahy and Call 2006, Zentall 2006b, Raby et al. 2007). As mentioned above, these ideas can be traced back to an earlier cognitive psychology, flourishing around 1920–1950 (Thorpe 1963). For example, Köhler (1924) observed chimpanzees solving problems, like reaching for a suspended banana by moving a box under it. When the solution came after a period of inactivity, Köhler inferred that the animal had been reasoning about possible courses of action until it arrived at an "insight" and formulated a suitable plan. Blodgett (1929) and Tolman and Honzik (1930b) introduced the idea of "latent learning" (learning without reinforcement) and Tolman (1948) the idea of the "cognitive map." The hallmark of these early cognitive theories, as well as of modern ones, is that experiences are represented in memory in a flexible form that can be used later to solve new problems, through a process similar to human thinking. The existence of these representations and processes in animals is the subject of enduring debate (Washburn 1908, Macphail 1982, Bekoff et al. 2002, Shettleworth 2010a, Heyes 2012a, Haselgrove 2016, Redshaw et al. 2017, Colombo and Scarf 2020). Three aspects of this debate are important in this book.

First, thinking-like mechanisms are implicitly assumed to be superior to associative learning, but this is not necessarily true once we acknowledge that thinking also has costs. For example, a cognitive map may take too long to build to be worthwhile, especially if the problems it excels at solving do not arise often. This point is discussed in chapter 11. Second, associative learning is often dismissed without proof or based on the simplest stimulus-response models. Associative learning, however, is more powerful than usually acknowledged. For example, it can result in the kind of sudden progress toward the solution of a problem that is usually interpreted as "insight." We reprise this point in the next section, and more extensively in chapters 3 and 7.

Our third point subsumes the first two, and concerns the need for theoretical understanding. Evaluating a putative information-processing mechanism requires formulating it mathematically, so that it can be studied rigorously. This is possible for many associative learning models (discussed in chapters 3–8). However, many "cognitive" ideas about animal intelligence have not been sufficiently formalized and remain too slippery to evaluate from a mechanistic, functional, and evolutionary standpoint.

2.2.3 Reinforcement Learning

Reinforcement learning models are artificial intelligence systems that improve autonomously from experience, based on receiving "reinforcement"—that is, signals of good or poor performance (Wiering and van

Otterlo 2012, Sutton and Barto 2018). For example, a reinforcement learning agent playing a video game of basketball might receive the number of points scored as reinforcement. As the name suggests, these models were initially inspired by animal psychology (Sutton and Barto 1981, Shah 2012) but were later developed with little contact with psychology and biology. Gallistel (1999) provides a nontechnical summary, Sutton and Barto (2018) a comprehensive treatment.

Reinforcement learning has the potential to change our understanding of animal intelligence dramatically, by providing mathematical models of how animals can learn sequences of behavior, become goal seeking, and make decisions based on what they have learned. For example, we can use reinforcement learning to formalize associative learning mathematically and show that it can learn the same behavior as thinking-like processes (Sutton and Barto 2018, Enquist et al. 2016). When applied to neural networks, reinforcement learning can also generate insights into perceptual learning and generalization. Reinforcement learning can also benefit the study of "more cognitive" hypotheses of animal intelligence, such as planning based on cognitive maps. These hypotheses, in fact, can also be formalized as reinforcement learning models (Sutton and Barto 2018), which is helpful to test them rigorously (see chapter 11).

Their merits notwithstanding, reinforcement learning models were not developed with animal learning in mind. They must be adapted and complemented to include, for example, genetic predispositions and motivational systems. Moreover, they typically operate under different conditions from animals. It is common for reinforcement learning models to receive millions of learning experiences, and to perform very badly initially. Animals, on the other hand, must learn rapidly, and must perform well enough to survive as they learn. Some robotics scenarios are closer to the challenges of animal life, but, in general, the best technical solution for an artificial intelligence task is not necessarily the most accurate model of animal intelligence. Chapters 4 and 5 deal in part with adapting reinforcement learning models to the study of animal intelligence.

2.3 Our View of Animal Intelligence

We believe that ethology, animal psychology, and cognitive approaches all contain crucial insights about animal intelligence, which can converge into a single theory. Several valuable attempts have been made in this direction, but, so far, no consensus has been reached (Hinde 1970, Fantino and Logan 1979, McFarland 1985, Timberlake 1994, Timberlake and Silva 1995, Reznikova 2007, Shettleworth 2010b, Hogan 2017). We suggest that an effective synthesis is now possible by integrating knowledge using the mathematical framework of reinforcement learning. We outline our attempt at such a synthesis in chapters 3–8, which introduce and discuss the A-learning model of animal intelligence (Enquist et al. 2016, Lind et al. 2019, Ghirlanda et al. 2020), a development of the Rescorla and Wagner (1972) model. There are three crucial elements in our theory.

The first one is that it integrates learning, decision making, and memory into a single machinery. This is necessary in a complete theory of animal intelligence, as these three aspects of information processing operate together. Learning uses experiences to modify memory, memory provides information for decision making, and decision making generates behavior, which in turn results in new experiences. A-learning features two learning processes that modify two kinds of memory variables: estimated values of stimuli and estimated values of actions. A decision-making process then uses estimated action values to generate actions. All three of these processes are assumed to be domain- and species-general, that is, to exist across species (at least vertebrates) and to be used across behavioral contexts, such as in foraging, social learning, and predator recognition.

The second element of our theory allows for domain- and species-specific adaptations, which are created by genetic predispositions that favor some behavioral outcomes over others. In this way, specific behavioral adaptations can emerge from general learning and decision-making processes. Genetic predispositions can operate on many information-processing elements, including perception, memory, learning, decision making, motor patterns, and motivational control (see chapter 4). Perhaps ethology rejected the idea of general learning processes in part because, at the time, it was unknown how such processes could be tuned genetically to produce species-specific adaptations.

The third element of our theory is that it is formalized mathematically, including the domain-general processes and their tuning through genetic predispositions. This is a big advantage for evaluating the theory. Using this formalization, we can simulate different learning scenarios to derive rigorous predictions and check their consistency with data. We can also explore how genetic predispositions can influence learning. For these purposes, we use the learning simulator mentioned in section 1.5.5.

We acknowledge that the A-learning model discussed in this book is not yet a complete theory of animal behavior. For example, it does not include perceptual learning and motivational processes—in part because we have not completed their formalization and in part because these aspects are less relevant to the presentation of our theory of the human evolutionary transition.

2.4 Human Intelligence

As summarized in section 1.1, many unique traits appeared during human evolution, and many of these quite suddenly, on the timescale of genetic evolution. We refer to these sudden changes as the human evolutionary transition. This characterization reflects our belief that human evolution represents a new strategy for discovering productive sequences of behavior and keeping combinatorial dilemmas in check (hypothesis 1 above). It is a transition from animal-like information processing, based on associative learning and genetic predispositions (hypotheses 2 and 3), to a new kind of information processing based on mental flexibility and cultural information. Our analyses of information processing and combinatorial dilemmas suggest the following hypotheses:

Hypothesis 4: Human intelligence relies on collecting large amounts of information, possibly of uncertain value, and on reorganizing this information internally to discover new productive behavioral sequences. Combinatorial dilemmas arising from this strategy are managed using cultural information to guide the search for productive sequences, rather than through genetic predispositions.

Hypothesis 5: The transition to this new information-processing strategy depends on relatively few changes to domain-general inborn abilities. At a minimum, these changes include enhanced abilities to represent sequential information and increased mental flexibility, enabling humans to learn new ways to process information.

Hypothesis 6: These changes to inborn abilities, while few, were decisive in enabling cultural evolution to an extent not seen in animals. After its onset, cultural evolution transformed human evolution and was instrumental in creating the most distinctive aspects of human intelligence.

Hypothesis 7: The origin of the human evolutionary transition lies in life history changes that substantially increased the duration of childhood and the rate of communication between children and adults, thereby starting a positive feedback between the transfer of cultural knowledge and the production of such knowledge. Longer childhood promoted cultural transmission from adults to children, which in turn resulted in children learning the mental skills to create more culture.

We begin discussing these hypotheses in chapter 9, which details what we consider uniquely human traits, and thus what we think must be explained. The chapter also contains an overview of major theories of human evolution. In chapter 10, we spell out hypotheses 4–7 in more detail, paving the way for an evolutionary analysis of human intelligence. Chapter 11 compares animal-like and humanlike information processing to delineate the advantages and disadvantages of each. We call humanlike information processing *thinking*, but we do not rule out a priori that animals can think, as discussed in section 2.5. Chapters 12–14 discuss the social transmission of behavioral sequences, the social transmission of mental skills, and human cooperation, respectively. These topics are all informed by our analysis of combinatorial dilemmas. Chapter 15 shows that, for a species that thinks, cultural evolution is essential to overcome combinatorial dilemmas. Finally, chapter 16 addresses the origins of the human evolutionary transition.

2.5 Thinking

In this section, we clarify our definition of *thinking*, which hypothesis 4 refers to. By *thinking*, we mean mental mechanisms that, at the time of decision making, can examine and recombine information that has been gathered previously, possibly in many different situations. Thus, in thinking mechanisms, the use of information is distinct from the gathering of information. Let us clarify with an example.

Consider an animal seeing an unknown fruit at location A and deciding *not* to eat it. Later on, at another location, the animal sees an identical fruit. This time, it decides to eat the fruit and finds it sweet. Will the animal go back to A to eat the first fruit? We can analyze the situation as follows. The first experience can be written as

$$\text{Go to A} \rightarrow \textit{Fruit} \tag{2.1}$$

That is, going to A was followed by seeing the fruit (we write actions in roman, stimuli in italics). The second experience can be summarized as

$$\textit{Fruit} \rightarrow \text{Eat} \rightarrow \textit{Sweet} \tag{2.2}$$

The crucial aspect of this example is that the value of the first experience becomes clear only at the time of the second experience. Most animals would learn from the second experience that eating the fruit is profitable, but which animals would realize that going back to A is profitable? Going to A has never been followed by a positive consequence, and concluding that it is profitable requires at least two steps:

1. The animal must *remember* the first experience at least until the time of the second experience. Thus the information that a fruit is at A must be memorized even if its value is unknown. It is not enough to remember "I saw a fruit somewhere" or "going to A produced something

unexpected." Both causes (Go to A) and effects (*Fruit*) must be remembered in some detail.

2. The animal must *recombine* the two experiences. By *recombine*, we mean that information from the two experiences must be considered jointly in order to draw the conclusion that going to A is profitable. Each experience is, on its own, insufficient to draw the conclusion.

It is the long-term storage of information of uncertain value, its subsequent recall, and the joint consideration of information acquired at different times that characterizes what we call *thinking*. We also stress that the ability to accurately represent and process sequential information is critical to thinking (see chapter 11). For example, it would be worthless remembering the sequence "Go to A → *Fruit*" in the opposite order, or to formulate a scrambled plan like "Eat → *Fruit* → Go to A." Indeed, we can characterize thinking as the ability of *mental sequencing*, that is, the ability to mentally discover new, productive sequences of actions by combining information from disparate sources.

We find this characterization of thinking appealing for three reasons. First, it fits well with what people seem to do when they think, reason, or plan, and with how these terms are used in animal and human cognition research. Second, it does not specify exactly how thinking works, only what it can accomplish. More specifically, our definition does not hinge on any capacity that is uniquely human, such as formal logic or a language with complex grammar. By design, it leaves open the possibility that animals can think, either similarly to humans or using different mental mechanisms. Third, the characterization stresses that thinking has evolved to find productive sequences in ways that are inaccessible to nonthinking mental mechanisms. Thus we can gain insight into its evolution by analyzing the conditions under which thinking is more productive than preexisting mental mechanisms, such as associative learning. Indeed, we see in chapter 11 that thinking is not always superior to associative learning.

While we eventually argue that only humans "think" among animals (chapters 7 and 8), we carefully qualify this statement in several ways. First, our conclusion comes from analyzing empirical data and not from prejudice. We are open to data showing that animals think. Second, stating that animals do not think *according to our definition* is not the same as saying that animals are not smart, cognitive, or intelligent, or that they do not think according to other definitions. Our statement only indicates that, after looking at available data, we do not believe that animal intelligence fulfills requirements 1 and 2 above, while human intelligence does. Finally, that humans can think does not imply that they think all the time, that our thinking is perfect, or that other mental mechanisms, such as associative learning, have become unimportant to human behavior (Sun 2001, Kahneman 2011).

2.6 Human Evolution

Human evolution—genetic and cultural—has been the subject of much research, yet little consensus exists about the causes of the human evolutionary transition. We discuss proposed explanations in some detail in section 9.5. Broadly speaking, these vary along two dimensions: the uniquely human mental skills to be explained and the roles of genetic and cultural evolution in shaping these skills.

Fundamentally, theory of human evolution should explain why humans became different from other animals. Many theories hypothesize that the genetic evolution of new mental skills was driven by a change to the physical or social environment. The venerable savanna hypothesis, for example, holds that human intelligence—together with physical traits, such as bipedalism and hand dexterity—evolved as a consequence of human ancestors moving from forests to the savanna, described as a harsher environment with fiercer competition for resources (Dart 1925). One difficulty with this and other hypotheses based on environmental change is that many species experienced the same change yet did not evolve noticeably higher intelligence. Environmental factors can be at most a partial cause of the human evolutionary transition.

A theory of human evolution should also explain archaeological and historical patterns. For instance, it must explain a large number of human phenomena that appeared too recently, and too quickly, to be the result of genetic evolution. These phenomena include practical skills, such as weaving, archery, or sailing, as well as mental skills, like arithmetic, reading, or playing chess. All these indicate exceptional motor and mental flexibility, leading to the ability to learn a large number of skills without specific genetic support. As seen in section 1.3, both mental and motor flexibility can be disadvantageous due to combinatorial dilemmas; thus it is crucial to explain how these dilemmas have been managed in human evolution. We see in chapter 9 that some theories of human evolution struggle with explaining human flexibility and its manifestation in mental and practical skills of recent invention. The main reason is that cultural evolution plays a minimal role in these theories. In particular, the idea that cultural evolution can result in new mental skills that are transmitted between individuals is rarely discussed (Parker 1978, Gabora 1997, 1998). These topics are central to chapters 9, 13, and 15.

Finally, a theory of human evolution should be compatible with findings about human development from different fields, such as developmental psychology, education science, and developmental linguistics. These findings show that many uniquely human skills, including mental skills, develop gradually in individuals, and necessitate both substantial individual effort and social input, often in the form of explicit tutoring (see chapter 9).

What would a theory that can contribute to explaining all these interconnected aspects of human evolution look like? With some important exceptions, most theorizing about human evolution has been based on verbal arguments. These, however, are insufficient for understanding long chains of causes and effects of the complexity found in human evolution. In this book, we advocate using an approach similar to evolutionary biology and behavioral ecology, which rests on detailed analysis of mathematical models of possible evolutionary scenarios (Krebs and Davies 1978, Houston and McNamara 1999, Westneat and Fox 2010). In the case of human evolution, this analysis requires tying together mathematical models of mental mechanisms, ecological factors, and genetic and cultural evolution. A sketch of such an analysis is presented in chapters 13–16, and its results are summarized next.

2.7 Our View of Human Evolution

The human evolution scenario explored in this book can be summarized as follows. The human evolutionary transition was initially driven by genetic changes to domain-general learning and memory abilities (Parker 1978), rather than by changes in specific domains, such as in linguistic abilities or theory of mind. A comparison with animals suggests that these genetic changes enhanced human abilities to represent, store, and process sequential information. These enhanced abilities started to emerge through genetic evolution at the beginning of the human evolutionary transition and ushered in a period of gene-culture coevolution. During this phase, the human social environment grew richer in information that could be learned by children, and humans became less reliant on genetic information in order to learn productive behavioral sequences. The progressive removal of genetic constraints on behavioral and mental functioning resulted in increased flexibility and sparked cultural evolution. The latter completed the shift to a new evolutionary strategy, in which culture, rather than genes, provides the necessary information to manage combinatorial dilemmas. Reliance on cultural evolution, in turn, can explain the great diversity in human mental and behavioral skills, including skills that are too recent to have evolved genetically, as well as the prodigious demographic expansion of humans to almost all terrestrial habitats. These ideas are developed starting with chapter 9. In the following, we offer some preliminary observations.

2.7.1 Cultural Evolution

We suggest that cultural evolution has had a larger scope in human evolution than often thought. The power of genetic evolution in crafting astonishing adaptations is recognized as deriving from the three properties of mutation, transmission, and selection of information. Culture is a similar system, based on social rather than genetic information, which can also have far-reaching consequences (Cavalli Sforza and Feldman 1981, Boyd and Richerson 1985, Richerson and Boyd 2005, Henrich 2015, Heyes 2018). We suggest, in fact, that the initial genetic changes to human behavioral and mental mechanisms did not immediately make humans much different from animals. Rather, their effect was to make cultural evolution possible. Only as cultural evolution picked up (which may have required hundreds of thousands of years; Lind et al. 2013) did humans become the species we know today, with unique skills in, for example, thinking and social learning. Of note, we propose that cultural evolution is not primarily about gathering factual knowledge and behavioral skills, such as in food processing, hunting, or medicine. While such knowledge is certainly important in fueling human evolution, we think that the most profound effect of cultural evolution has been to create new mental skills, such as skills that foster creativity, planning, social learning, and other uniquely human traits (Heyes 2018, Moore 2020).

2.7.2 Genetic Evolution

While we emphasize cultural evolution, we do not deny that genetic evolution played a major role in the human evolutionary transition. What we propose is that genetic changes have affected mostly domain-general processes that did not immediately result in overtly unique skills (Sloutsky 2010, Barry et al. 2015). As mentioned above, we suggest that genetic changes in human memory resulted in a better representation of sequential information, and that the human mental and behavioral machinery became more flexible. Most brain functions, however, were not affected by these changes, and humans continue to share mental mechanisms, such as associative learning and motivational systems, with animals.

We do not deny that specific abilities, such as grammatical competence, are supported by specific genetic factors, but we think that even specific abilities must leverage uniquely human domain-general mental mechanisms. The main reason for this proposal is that sequential information processing is crucial for all uniquely human mental and behavioral abilities (see chapter 9); thus we deem it more likely that abilities for such processing have evolved once rather than multiple times.

In summary, we suggest that the role of genetic evolution in the human evolutionary transition has been to make humans much more flexible than other animals. To use a technological analogy, we could say that animal mental mechanisms are similar to hardwired computers. They carry out efficiently the tasks they are built for but cannot be expanded at will with new software. The human mind, however, is more similar to a general-purpose computer whose capabilities can be greatly expanded through a staggering diversity of "cultural software."

2.7.3 What Started the Transition?

Our analysis of combinatorial dilemmas, in chapter 1 and later in the book, suggests that there are substantial hurdles to the evolution of humanlike intelligence, which could explain why it has evolved only once. For example, a "thinking" mental mechanism (section 2.5) must first learn a large amount of information before it offers any advantage over associative learning, which is more economical in resources and can be readily tuned to specific demands by genetic predispositions. If such information is not easily available, thinking may offer no advantages at all. In chapter 16, we look at this issue and related hurdles for clues about the start of the transition.

In short, we propose that the genetic evolution of general abilities for sequential information processing was triggered by decreased learning costs and increased access to social information. These changes were rooted in life history changes that occurred in our prehistory, including an increase in the length of childhood that gave children more time to learn while cared for, and an increase in the rate and quality of adult-child interactions that made more information available to children (Kaplan et al. 2000, 2003, Konner et al. 2010, Gopnik et al. 2020). Thus, compared to other theories, our proposal puts less emphasis on environmental change, and rather suggests that domain-general thinking abilities were genetically selected for once the quantity and quality of adult-child interactions created an information-rich environment that made thinking a worthwhile evolutionary strategy.

3

Learning Behavioral Sequences

- Behavioral sequences can be learned through associative learning in a process known as chaining.
- Chaining is powerful. Given time and information, it can learn any sequence.
- Chaining is computationally simple. Our model of chaining, A-learning, includes two equations to describe learning and one equation for decision making.
- Chaining is evolutionarily plausible. It can evolve easily from learning mechanisms that can learn only single behaviors.
- Chaining can account for how animals learn behavioral sequences, both when they succeed and when they fail.

3.1 Introduction

In this chapter, we discuss a simple yet powerful mental mechanism that may account for how animals learn most, if not all, behavioral sequences. We refer to this mechanism as *chaining*, a term introduced by Skinner to describe a type of learning in which known behaviors or shorter sequences are linked to form new or longer sequences (Skinner 1938, Kelleher and Gollub 1962, Mackintosh 1974, Williams 1994). Chaining combines learning about the value of actions (instrumental learning) with learning about the value of stimuli. For example, a dog that repeatedly hears a click before receiving food will eventually consider the click rewarding in itself, and will learn to perform behaviors that produce the click. The click is called a *conditioned* reinforcer because its reinforcing effects depend on a history of experiences, rather than being innate. In this chapter, we begin with a short introduction to modeling learning and behavior, and then present our model of chaining, called A-learning (Enquist et al. 2016, Ghirlanda et al. 2020). We then relate the model to observations of learning in nature and in the laboratory, with an emphasis on phenomena that are usually interpreted as evidence of mental mechanisms other than associative learning.

The model of chaining presented below forms the backbone of our analysis of animal intelligence. (Humans can learn by chaining, too, but our discussion of human cognition focuses on the differences between human and animal learning.) Chapter 4 considers genetic predispositions that influence learning and behavior, and explores how these can be integrated with chaining. Chapter 5 discusses what information animals use to learn and make decisions. Chapters 6 and 7 cover phenomena at the center stage of the debate on animal cognition, such as social learning, mental recombination of information, and planning. Chapter 8, our final chapter about animal intelligence, asks whether chaining can really be a complete model of animal learning.

3.2 Modeling Learning and Behavior

Behavior theory aims to understand the mechanisms that underlie behavior: how they operate, how they develop during life based on genetic and environmental influences, and how they evolve (Tinbergen 1963, Hogan 2017). As discussed in chapter 1, this book is primarily concerned with how animals learn sequences of behavior that they do not know innately (or that they know only partly, as discussed in chapter 4). We use the following notation to describe sequences of stimuli and behaviors:

$$S_1 \to B_1 \to S_2 \to B_2 \to S_3 \to B_3 \to \dots \qquad (3.1)$$

where S_1 is the first stimulus considered, B_1 is the behavior that follows S_1, S_2 is the second stimulus, and so on. To illustrate how we use equation 3.1, let's begin by considering a bird confronted with a ripe fruit. A simple description of this situation may include the stimuli and behaviors indicated in table 3.1.

Table 3.1. A few stimuli (*italics*) a bird may perceive in the presence of a fruit and a few behaviors (roman) it can enact

Event	Description
Fruit	Visual and olfactory perception of a fruit
Open fruit	Perception of an open fruit
Sweet	Sensations from licking the fruit's juice
Peck	Strike at fruit with closed beak
Lick	Lick the fruit
Touch	Touch the fruit with closed beak

Given that the bird is hungry, the best way to behave is to first peck the fruit and then lick its sweet juice, which we represent as

$$\textit{Fruit} \to \text{Peck} \to \textit{Open fruit} \to \text{Lick} \to \textit{Open fruit} + \textit{Sweet} \tag{3.2}$$

where stimuli are in italics and behaviors in roman. However, an inexperienced bird may engage in less efficient sequences, such as

$$\begin{aligned} &\textit{Fruit} \to \text{Lick} \to \textit{Fruit} \to \text{Touch} \to \Big[\textit{Fruit} \to \text{Peck} \to \textit{Open fruit} \to \\ &\quad \to \text{Lick} \to \textit{Open fruit} + \textit{Sweet}\Big] \to \text{Peck} \to \textit{Open fruit} \end{aligned} \tag{3.3}$$

In this case, the rewarded sequence (between brackets) occurs as part of a longer sequence. Thus (3.2) represents the optimal behavior, while (3.3) represents an actual behavioral sequence that departs from the optimal one. An alternative notation that we sometimes use to display longer sequences is as follows:

$$\begin{array}{lcccc} & \text{Stimulus:} & & \text{Behavior:} & \\ \text{Step 1:} & \textit{Fruit} & \to & \text{Peck} & \to \\ \text{Step 2:} & \textit{Open fruit} & \to & \text{Lick} & \to \\ \text{Step 3:} & \textit{Open fruit} + \textit{Sweet} & & & \end{array} \tag{3.4}$$

We stress that these notations simply represent a succession of events. They do not imply that a behavior depends only on the preceding stimulus. For example, an individual's sexual behavior can depend both on current stimuli (how a potential partner looks and behaves) and on stimuli experienced much earlier in life (how the individual's parents looked and behaved; Bischof 1994). Nevertheless, we can often restrict our analysis to stimuli and behaviors that occur in a particular context, rather than considering an animal's whole lifetime. To study how a bird learns to eat fruit, for example, we would consider only stimuli and behaviors that occur when the bird is in the presence of fruit.

We recognize that representing sequences as in equation 3.1 or equation 3.4 is a simplification. Sensory input and behavioral output occur in real time and are not neatly arranged in an alternating sequence. At the same time, our notation conveniently represents the observation that behavior is commonly triggered by specific stimuli, both in the wild and in the laboratory. The analysis of stimulus and behavior sequences has a long history in experimental psychology (Skinner 1938, Fantino and Logan 1979, Pierce and Cheney 2018).

Behaviors and stimuli can be described in different levels of detail, with the appropriate amount of detail dependent upon what we wish to understand.

To understand motor coordination, for example, behavior must be described in terms of muscle contractions. In this book, however, we use high-level behavioral categories that indicate what a behavior accomplishes, such as "Peck" or "Lick" (Pierce and Cheney 2018). Similarly, we typically describe stimuli in terms of a few relevant features of the stimulus, such as "*Fruit*" or "*Open fruit*" in table 3.1, and we assume that animals have access to this information. Understanding how animals actually recognize stimuli requires a more detailed analysis that is beyond our present scope (Arbib 2003, Enquist and Ghirlanda 2005). We also sometimes consider stimuli that originate within the animal, such as hunger or thirst.

As a framework for model building, we consider behavior systems as dynamical systems in which specified rules determine the system's output (behavior) given the input (stimuli) and the system's internal states, such as memories and motivations (McFarland 1971, Luenberger 1979, Houston and McNamara 1999). To build such a model, we need to specify what variables enter the model, how these variables determine behavior, and how they change as a consequence of experience. In other words, we need to consider the following elements:

Information: What information is used by learning and behavior mechanisms? External and internal stimuli are obvious sources of information. Another source of information is *memory*, whereby experiences can cause changes in the nervous system that can affect future behavior. One of our goals is to understand what information animals and humans store in memory.

Learning: How is memory updated based on experience? As discussed in section 1.3.3, learning must be selective—remembering everything that ever happens is not an effective strategy. What information is stored and how it is updated are crucial elements of any model of learning.

Decision making: How are stimuli and memories used to select behavior? One goal of decision making is to exploit knowledge that the animal possesses. For example, given that an animal is hungry, decision making should select behaviors that are known to yield food. Decision making must also support exploration. That is, behaviors with unknown or uncertain outcomes should sometimes be selected so that their utility can be evaluated (also known as the exploration-exploitation dilemma; Sutton and Barto 2018).

In the rest of this chapter, we develop a model of sequence learning based on specific assumptions in these three areas. We refine the model in the following chapters.

3.3 Learning a Single Behavior

Before introducing our model of sequence learning, let's consider how an animal may learn a single behavior. The core concept is to use exploration to estimate the value of different behaviors and then to select behaviors with higher values (Sutton and Barto 2018). For example, we may consider a short sequence like

$$\textit{Light} \rightarrow \text{Press lever} \rightarrow \textit{Food} \tag{3.5}$$

as is frequently encountered in laboratory experiments. The first stimulus, *Light,* is initially meaningless to the animal. Learning consists of the discovery that in the presence of light, pressing the lever leads to food. Equation 3.5 is a particular instance of a general sequence with the form

$$S \rightarrow B \rightarrow S' \tag{3.6}$$

where S' is a stimulus with *primary value,* often referred to as a *reinforcer*. The function of reinforcers is to guide learning toward adaptive outcomes. We assume that primary values have been set by genetic evolution to reflect how a stimulus can contribute to the animal's survival and reproduction. For example, we may assume that painful stimuli have negative value because pain is a signal of damage to the body, and that food ingestion has a positive value to a hungry animal. Stimuli with negative value should be avoided, and those with positive value should be sought. We use the notation $u(S)$ to indicate the primary value of S. For example, in equation 3.5 we may assume $u(\textit{Light}) = 0$ because the animal has no innate knowledge about the light, while we assume $u(\textit{Food}) > 0$, the exact value dependent upon the food. We now make some formal assumptions about information, learning, and decision making.

Our initial (simplified) assumption about information is that stimulus S contains all information necessary to behave optimally, and that this information is readily perceived by the animal. In reality, perceived stimuli can be more or less informative about the state of the environment. For example, a gazelle spotting a lion will not immediately know whether the lion is hungry or not. Stimulus perception and recognition are important problems (see chapters 5 and 8) but not central to our arguments.

Regarding learning, we assume that the goal of the learning mechanism in any given situation is to estimate the value of behaving in a certain way. Let $v(S \rightarrow B)$ be the estimated value of performing behavior B when encountering stimulus S. We refer to $v(S \rightarrow B)$ as a *stimulus-response value*. This value is the model's memory. Assume now that S is perceived, B is chosen, and S' is experienced, as in equation 3.6. If S' has value $u(S')$, the correct value for $v(S \rightarrow B)$ is $u(S')$, because this is what the animal gains when responding with

B to stimulus S. Thus we would like the model to learn that $v(S \to B) = u(S')$. This can be accomplished by introducing a positive "learning rate" parameter, α_v, and updating $v(S \to B)$ according to

$$\Delta v(S \to B) = \alpha_v [u(S') - v(S \to B)] \qquad (3.7)$$

where $\Delta v(S \to B)$ is the change in $v(S \to B)$ caused by the experience. According to equation 3.7, $v(S \to B)$ changes in proportion to the difference between its current value and the correct value, $u(S')$. If $v(S \to B)$ is lower than $u(S')$, $\Delta v(S \to B)$ will be positive and $v(S \to B)$ will increase. Conversely, if $v(S \to B)$ is higher than $u(S')$, $\Delta v(S \to B)$ will be negative and $v(S \to B)$ will decrease. Eventually, $v(S \to B)$ will approximate $u(S')$ closely, provided α_v is small enough (Widrow and Stearns 1985, Haykin 2008). Many readers will have recognized in equation 3.7 the classic equation of associative learning by Rescorla and Wagner (1972) . The only difference is that we refer to value rather than associative strength in order to highlight the biological function of learning.

We now consider decision making. It seems obvious that behaviors with higher value should be chosen more often. However, decision making should also leave room for exploring other behaviors. This is important both in the initial stages of learning, when stimulus-response values can be inaccurate because of inexperience, and later on, should a behavior change in value. A simple decision-making function that preferentially selects high-value behaviors while leaving room for exploration is

$$\Pr(S \to B) = \frac{e^{\beta v(S \to B)}}{\sum_{B'} e^{\beta v(S \to B')}} \qquad (3.8)$$

where $\Pr(S \to B)$ is the probability that B is chosen when perceiving stimulus S, β is a positive parameter, and the sum includes all possible behaviors. To see how equation 3.8 works, note first that $e^{\beta v(S \to B)}$ is an increasing function of $v(S \to B)$. Hence the probability of each behavior increases with its stimulus-response value. The parameter β determines how quickly $e^{\beta v(S \to B)}$ increases with $v(S \to B)$. Its role is to regulate exploration: if β is small, even behaviors with low positive value are selected with appreciable probability, whereas if β is large, only behaviors with the highest value occur with any likelihood. Finally, the sum within the denominator ensures that the probabilities for all behaviors sum up to 1. While we do not claim that equation 3.8 is exactly the mechanism used by animals, it is broadly compatible with known aspects of choice behavior. For example, the relative probability of choosing two alternative behaviors is predicted to depend on the difference between the

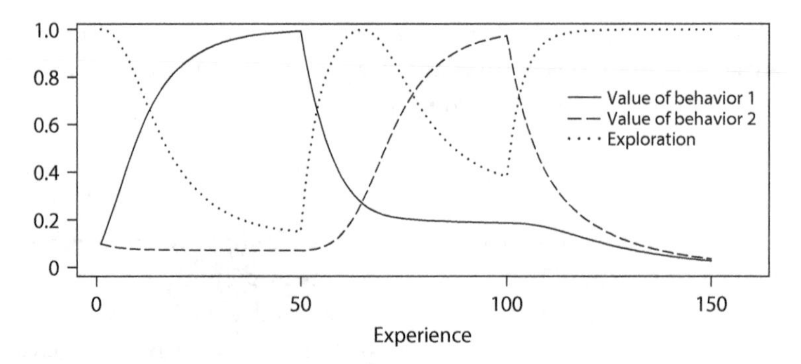

FIGURE 3.1. Expected course of learning about the value of two actions, according to equation 3.7. The model repeatedly experiences the same environmental state (stimulus). For the first 50 experiences, action 1 is valuable; then action 2 becomes valuable for a further 50 experiences; and, finally, neither action is valuable. The learning rate is $\alpha_v = 0.1$ and behaviors are chosen according to equation 3.8 with $\beta = 5$. The dotted line represents the amount of exploration, defined as $1 - \sigma^2/\sigma^2_{\max}$, where σ^2 is the variance in choice of behavior and $\sigma^2_{\max}$ the maximum possible variance. Given m behaviors with probabilities p_i, the variance is $\sigma^2 = \frac{1}{m}\sum_{i=1}^{m}(p_i - \frac{1}{m})^2$ while the maximum variance is $\sigma^2_{\max} = \frac{m-1}{m^2}$, obtained when one of the p_i's is 1 and the others 0. See section 1.5.5 for simulation details.

values of the behaviors:

$$\frac{\Pr(S \to B_1)}{\Pr(S \to B_2)} = \frac{e^{\beta v(S \to B_1)}}{e^{\beta v(S \to B_2)}} = e^{\beta(v(S \to B_1) - v(S \to B_2))} \tag{3.9}$$

which is compatible with the matching law of experimental psychology, stating that the probability of a response is an increasing function of reward that can be gained from the response (Herrnstein 1974, Baum 1974, Ghirlanda et al. 2020; see Budaev et al. 2019, for a discussion of animal decision making).

Now that we have one equation for learning and one for decision making, we can study how the model learns while it explores its environment. To demonstrate the flexibility of the model, we start from the example in equation 3.5 but add a second behavior, say, "Pull chain." We assume that "Press lever" is rewarded during an initial time period, that "Pull chain" is rewarded during a second period, and that neither behavior is rewarded during a third period. Figure 3.1 shows that the model successfully tracks these values. Of particular interest is the transition between the first and second periods. Just after this transition, the model is predominantly choosing "Press lever" because $v(\textit{Light} \to \text{Press lever})$ is high. This choice, however, no longer results in a reward, hence $v(\textit{Light} \to \text{Press lever})$ decreases. As this occurs, (3.8) automatically starts choosing "Pull chain" more often (shown by the dotted line as increased exploration), which enables the model to discover that this behavior is now valuable. Thus $v(Light \rightarrow \text{Pull chain})$ increases while $v(Light \rightarrow \text{Press lever})$ decreases, as appropriate. In summary, the model successfully tracks the value of actions and is capable of resuming exploration if an established behavior is no longer rewarding. With reference to our discussion of information recombination in section 2.5, we note that there is only very limited use of information in this model, in that decision making is based solely on past experiences with the current stimulus and no other stimuli.

3.4 Learning Behavioral Sequences through Chaining

The model in the preceding section cannot learn sequences of behaviors. To appreciate the problem, consider a task in which two behaviors have to be performed in sequence in order to reach a rewarding state:

$$S_1 \rightarrow B_1 \rightarrow S_2 \rightarrow B_2 \rightarrow S_3 \tag{3.10}$$

In this case, only S_3 has reward value. The model in the previous section can learn to respond with B_2 to S_2, because this response is followed by a rewarding stimulus. It cannot, however, learn to respond with B_1 to S_1, because the following stimulus, S_2, has no reward value. This impasse can be resolved by learning to attach value to S_2, since experiencing S_2 can lead to experiencing the valuable stimulus, S_3. If S_2 comes to have value, experiencing it can reinforce the use of B_1 in response to S_1, thus establishing the complete behavior sequence in equation 3.10. A stimulus that acquires value by virtue of predicting another valued stimulus is known in experimental psychology as a conditioned reinforcer (Williams 1994). We model conditioned reinforcement by assuming that any stimulus S can have both primary value, $u(S)$ as in equation 3.6, and *stimulus value*. This is a new quantity that we represent as $w(S)$. We modify the previous model by introducing stimulus value as follows. After the sequence $S \rightarrow B \rightarrow S'$ occurs, $w(S)$ is updated to

$$\Delta w(S) = \alpha_w[u(S') + w(S') - w(S)] \tag{3.11}$$

where α_w is a learning parameter similar to α_v in equation 3.6. This equation simply states that the stimulus value of S is updated to approximate the total value of the stimulus that is experienced next (compare equation 3.6), which in turn is the sum of the primary value of the stimulus, $u(S')$, and of its stimulus value, $w(S')$. We also modify equation 3.6 to include stimulus value:

$$\Delta v(S \rightarrow B) = \alpha_v[u(S') + w(S') - v(S \rightarrow B)] \tag{3.12}$$

This equation is similar to equation 3.7 but with a target value for $v(S \rightarrow B)$ of $u(S') + w(S')$ rather than of $u(S')$ alone. Note that stimulus values are

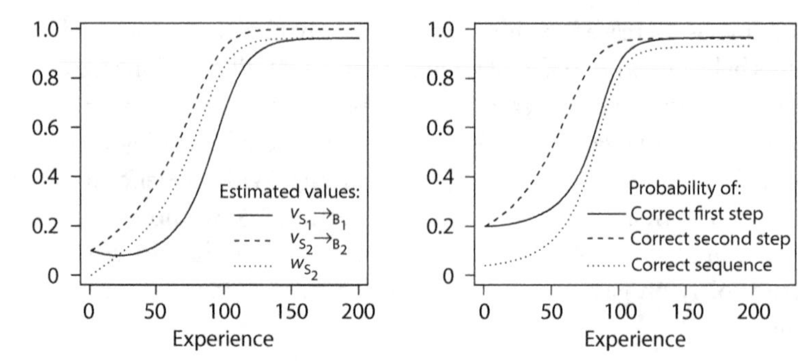

FIGURE 3.2. Learning a sequence of two behaviors using chaining. The correct sequence $S_1 \to B_1 \to S_2 \to B_2 \to S_3$ yields a reward of 1, and all other sequences are unrewarded. An "attempt" refers to all steps taken from the perception of S_1 until a mistake occurs or until S_3 is reached. The left panel shows the expected course of memory changes. The right panel shows the probability of choosing the correct response at the first and second steps and the overall probability of a correct sequence. The correct behaviors must be discovered among $m = 5$ possible behaviors. See section 1.5.5 for simulation details.

updated solely based on the succession of stimuli, regardless of what behavior is performed (the behavior B does not enter equation 3.11). In contrast, a stimulus-response value is specific to a behavior and is only updated when that behavior occurs.

Equations 3.8, 3.11, and 3.12 represent the formal specification of our chaining model. We have called it A-learning because of its focus on animal learning (Ghirlanda et al. 2020), but in this book we often refer to it simply as chaining. Equations 3.11 and 3.12 were first proposed by Wiering (2005) as a machine learning algorithm, which we have adapted to study animal learning and behavior.

From the point of view of information processing, chaining still performs trial-and-error learning with very limited integration of information. However, in contrast with the model in section 3.3, chaining can learn complex sequences of actions. For example, consider again the sequence in equation 3.10. The first time $S_2 \to B_2 \to S_3$ is experienced, the stimulus value $w(S_2)$ increases because $u(S_3)$ is positive. Once $w(S_2)$ is positive, the experience $S_1 \to B_1 \to S_2$ will be able to increase $v(S_1 \to B_1)$. Over many experiences, both $w(S_1)$ and $w(S_2)$ will approach $u(S_3)$ and thus reinforce the correct response to S_1 and S_2, as shown in figure 3.2.

The chaining model that we have just presented forms the backbone of our account of animal intelligence based on associative learning and genetic predispositions, which we continue to develop from here through chapter 8.

Some readers may wonder why we do not use the traditional concepts of Pavlovian and instrumental conditioning, and why our analysis of animal learning seems geared toward instrumental learning, even though Pavlovian conditioning is fundamental to behavioral adaptation (Roper 1983, Hollis 1997).

The answer to the first question is that learning theorists still debate how to best understand Pavlovian and instrumental conditioning (Pearce 2008, Bouton 2016). We prefer to use the well-defined concepts of stimulus-response values and stimulus values to leave no ambiguity about the learning processes we are considering. This is not to downplay the achievements of animal psychology, as the question is genuinely complex and has been scrutinized extensively by learning theorists (Mackintosh 1983b, Hall 2002, Holland 2008). We refer readers to Ghirlanda et al. (2020) for a detailed discussion of instrumental and Pavlovian conditioning in A-learning. In brief, A-learning models both Pavlovian and instrumental conditioning using stimulus-response values and stimulus values. The only difference is that, in Pavlovian conditioning, genetic predispositions more strongly constrain which responses are learned. For example, in Pavlov's classic experiments in which food followed a neutral stimulus, dogs always learned behaviors related to feeding, such as salivating or sniffing the stimulus. In instrumental conditioning, a diversity of behaviors can be learned, as genetic constraints are weaker (though not absent; see chapter 4). For example, in order to obtain food, a rat can learn to sit, turn around, run, press a lever, move to a specific location, and so on.

One important aspect of Pavlovian conditioning that is directly included in A-learning is learning to attach value to stimuli, which is referred to as second-order conditioning, evaluative conditioning, and conditioned reinforcement in animal psychology. Stimulus value learning is essential to learn behavioral sequences, as seen above, and to understand the outcomes of many experiments, as we discuss below and in chapter 7. In this sense, Pavlovian conditioning is very much present in this book. Pavlovian conditioning is also relevant to our arguments because it can be seen as gathering knowledge about the world, such as knowledge that particular stimuli predict food or other meaningful events. Potentially, this kind of knowledge could be used in reasoning and planning (Hull 1952, MacCorquodale and Meehl 1954). We consider the evidence that animals gather and use such knowledge in chapter 7.

3.5 What Can Chaining Accomplish?

Our model of chaining, A-learning, is one of many machine learning models based on the principle of temporal differences (Sutton and Barto 2018), which in our case amounts to comparing the value of stimuli and responses experienced at successive times (equations 3.11 and 3.12). These models can learn arbitrary sequences of behaviors, provided that two conditions are fulfilled (Sutton and Barto 2018, Bertsekas and Tsitsiklis 1996, Bertsekas 2012, Enquist et al. 2016). First, the model must have access to sufficient information. As mentioned above, in this chapter we assume that the perceived stimuli contain all necessary information to learn optimal behavior (but see section 3.7 for some exceptions). What information animals actually use is discussed in chapter 5. Second, the learning mechanism must experience each stimulus-response pair a sufficient number of times, so that the optimal response can be determined accurately. This second condition requires that the model explore alternative courses of action sufficiently.

The amount of time the chaining model actually needs to learn a sequence depends on the length of the sequence and on two additional features. The first is the entry pattern of the sequence: Does the learner always start from the first step, or can it sometimes start from later steps? The second is the exit pattern: What happens when the animal makes a mistake? The worst-case scenario is when a sequence is always entered from the first step, and mistakes always cause a return to the first step. In this case, chaining fares little better than brute-force exploration, in that the number of attempts to learn a sequence still grows exponentially with sequence length. In practice, this means that only short sequences can be learned (section 1.3.1). This worst-case scenario is not specific to chaining; any learning mechanism needs to find the reward at least once, which initially can happen only by chance unless the mechanism already has some information about the sequence. However, sequences can be much easier to learn in some scenarios. A squirrel learning to open nuts, for example, may sometimes find a nut with an already cracked shell (a favorable entry pattern), and will not have to start from scratch if a bite or prying movement fails (a favorable exit pattern). Figure 3.3 and table 3.2 show that even a small probability of entering sequences from intermediate steps can shorten learning time dramatically. Relevant prior knowledge can also considerably facilitate learning, as we discuss in chapter 4.

In summary, chaining is a computationally simple algorithm that can learn sequences of behavior. But is it a good model of how animals learn? In the remainder of this chapter, we show that chaining can account for a diversity of observations about animal learning. We recognize that the behavior we attribute to chaining could also stem from other learning mechanisms, and we discuss in later chapters (particularly chapter 7) how chaining can be distinguished from these other mechanisms.

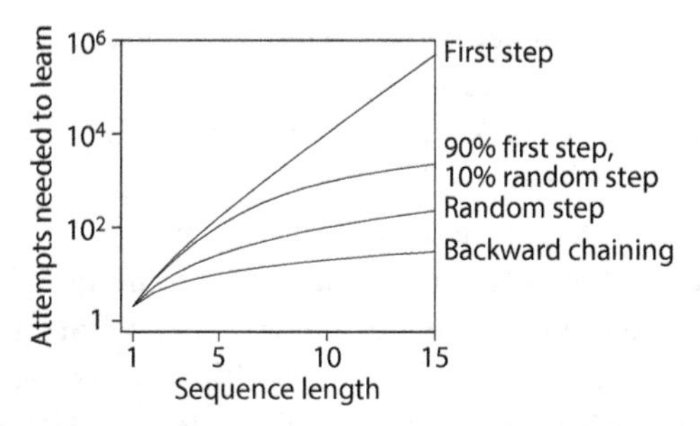

FIGURE 3.3. Effect of entry pattern on learning through chaining. The expected number of attempts needed to learn a behavior sequence is plotted against sequence length in four scenarios. From top to bottom, the sequence is entered from the first step; the first step 90% of the time, and a random step 10% of the time; a random step; and the highest step that has not been learned yet (backward chaining, section 3.3; forward chaining has the same expected learning time). A mistake terminates the current attempt. Note the logarithmic vertical axis. The analytical expressions for these curves are in table 3.2.

Table 3.2. Some ways in which an animal may learn a sequence of behaviors and their corresponding expected learning times

Learning method	Entry pattern	Exit pattern	Learning time
Animal on its own	Start	Back to start	Exponential $\sim am^l$
Animal on its own	Any state	Back to start	Quadratic $\sim \frac{am}{2} l^2$
Backward chaining	Highest state not yet learned	Stay in place	Linear $\sim alm$
Forward chaining	Start	Stay in place	Linear $\sim alm$

Note: Learning time given as a function of length of sequence (l), number of behaviors available to the animal at each step (m), and number of attempts required to learn one step in a given criterion (a). The latter can be decreased using shaping, which helps the animal discover the correct behavior. Forward chaining also includes the delivery of rewards upon reaching intermediate states.
Source: Calculations from Enquist et al. (2016).

3.6 Chaining in Animal Training

The most impressive examples of learned behavior sequences in animals come from professional animal trainers. We interviewed Bob Bailey, former general manager of Animal Behavior Enterprises, the first business ever to provide animal training services based on scientific understanding of behavior. He recounted having used chaining techniques with over 100 species, and that surprisingly long and complex sequences can be taught (see http://www3.uca.edu/iqzoo/iqzoo.htm for examples). Domestic cats, for example, can respond to a chain of verbal commands with near perfect performance for up to 40 minutes before receiving a reward, and dolphins can be trained to perform unassisted for hours. Backward chaining, forward chaining, shaping, and marking are the main techniques employed by animal trainers. These techniques are all derived from those employed by laboratory researchers in animal learning experiments. We describe these techniques briefly, focusing on the role of conditioned reinforcement and on how much they speed up learning, compared to an animal learning without a trainer by brute-force exploration. Table 3.2 summarizes this discussion. McGreevy and Boakes (2011) provide an excellent description of how these techniques are used in practice in animal training and husbandry. The reader should keep in mind that the traditional training terms we use here, forward chaining and backward chaining, are distinct from chaining as a theory of animal learning.

In *backward chaining*, we teach a sequence of behaviors starting from the last one and backtracking one step at a time. For example, suppose we are teaching a monkey to open a door with a key. We might break down the task as follows:

	Stimulus:		Behavior:	
Step 1:	*Key*	→	Grasp key	→
Step 2:	*Key in hand*	→	Insert key	→
Step 3:	*Key in lock*	→	Turn key	→
Step 4:	*Door unlocked*	→	Push door	→
Step 5:	*Reward*			

We assume that the monkey has already been taught the individual behaviors but has never put them together to open a locked door. To teach this sequence by backward chaining, we would first insert and turn the key ourselves, wait until the monkey pushes the door open, and then deliver a treat for reward. We repeat this step until it has been mastered. This establishes the link *Door unlocked* → Push door and endows the stimulus *Door unlocked* with stimulus value. We then train the second last step by inserting the key ourselves and waiting until the monkey turns it. When this happens, the monkey finds herself at the next step, *Door unlocked*. She now knows what to do and can complete the sequence. Importantly, the stimulus value of *Door unlocked* can reinforce the *Key in lock* → Turn key step, even if we do not deliver any reward when the monkey turns the key. We repeat this step until the last two steps are established, and then continue to train the preceding steps in the same way.

Backward chaining can be understood as the presentation of a specific entry pattern. During training, the animal always enters the task just one behavior away from a state it finds rewarding, because of either primary value or accrued stimulus value. Thus, to teach a sequence of l behaviors, we need

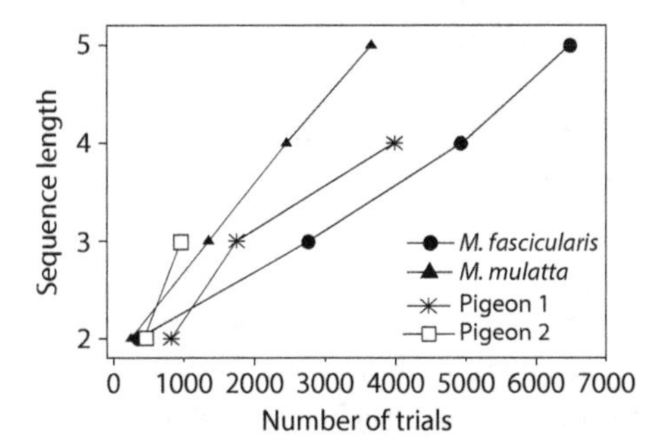

FIGURE 3.4. Number of attempts required to learn behavioral sequences of different length when trained by forward chaining. The observed linear relationships support the hypothesis that chaining is the mechanism underlying such learning (table 3.2). Learning criteria differed across studies: 80% correct for macaques (Colombo et al. 1993), 70% for pigeon 1 (Straub and Terrace 1981; only 5 of 7 pigeons learned the four-step sequence); and 75% for pigeon 2 (Terrace 1986). The sequence consisted of pressing response keys in a specific order.

only aml attempts, where a is the number of attempts needed to learn one behavior (assuming that all behaviors have equal difficulty) and m is the repertoire size. This learning time is merely linear with sequence length, while it would be exponential if we let the animal enter the task exclusively from the beginning (table 3.2).

Forward chaining works by manipulating the reward rather than the entry pattern. In our example, we start by rewarding the monkey for merely holding the key, then for holding it and inserting it, and so on. In terms of learning times, forward chaining has a benefit similar to that of backward chaining, because it also requires learning just one step at a time. Indeed, experiments using forward chaining to teach behavior sequences show that the time to master a sequence is proportional to the length of the sequence (figure 3.4). Note that forward chaining also relies on stimulus value. When we withdraw the reward for picking up the key, for example, the monkey is still rewarded for performing this step because it finds itself in the situation *Key in hand*, which has acquired stimulus value through a previous reward. Without conditioned reinforcement, forward chaining would not work, because intermediate behaviors would start to extinguish as soon as we stop rewarding them.

Shaping establishes a desired behavior by initially rewarding even rough approximations of it and then gradually restricting rewards to behaviors that increasingly resemble the desired one. Suppose we want to teach a pigeon to do a full turn. We may first reward a 45-degree turn when it occurs spontaneously. Once the pigeon has learned to turn 45 degrees, we stop rewarding 45-degree turns and set a new target at 90 degrees, and so on until the pigeon turns a full 360 degrees. For shaping to work, establishing an approximate behavior must increase the likelihood of an even better approximation. For example, a pigeon that has been rewarded for turning 45 degrees is more likely to turn 90 degrees than a naive pigeon. Shaping can be used to train single behaviors that can then be included in a sequence, as in our example of a monkey being trained to open a locked door.

Marking refers to yet another use of conditioned reinforcement. Before explicit training begins, it is common to establish as a conditioned reinforcer an unobtrusive stimulus, such as a click delivered before a food reward. After a few pairings of the click with the food, the click acquires value and can be used to guide the animal's behavior. In forward and backward chaining, for example, we can click when the animal behaves correctly at an intermediate step. This is referred to as "marking" the correct behavior. Marking adds stimulus value to intermediate steps, which makes them desirable to the animal.

The training techniques just discussed can be used in applied settings, but are also commonplace in research. For example, Inoue and Matsuzawa (2009) used forward chaining to teach chimpanzees to touch the numerals 1–9 in order. Table 3.3 lists a few studies employing backward or forward chaining. Without these techniques, teaching behavioral sequences would be extremely hard, if not impossible. For example, Straub et al. (1979) report that pigeons made little progress in learning to peck four images in a specific order before a tightly controlled forward chaining procedure was introduced (see also Straub and Terrace 1981, Terrace 1986, Pearce 2008).

3.7 Chaining in the Laboratory

Laboratory research shows that chaining is a powerful model that can account for a surprising diversity of findings about animal learning (Kelleher and Gollub 1962, Hendry 1969, Fantino and Logan 1979, Williams 1994, Ghirlanda et al. 2020). In this section, we consider two illustrative experiments.

In experimental psychology, an *observing response* is a response whose only outcome is to provide information about the environment. Figure 3.5

White → Peck key → *Food*
White → Peck key → *No food*
White → Press pedal → *Red* → Peck key → *Food*
White → Press pedal → *Green* → Peck key → *No food*

FIGURE 3.5. Event sequences in Wyckoff's (1952, 1969) experiment on observing responses. Other sequences that could occur (e.g., trials with neither key pecks nor pedal presses or with pedal presses only) are not relevant to our analysis and are omitted.

Table 3.3. Examples of trained behavioral sequences

Length	Species	Training method	Source
8–10	Dog	Backward chaining	McGreevy and Boakes (2011, 58)
12–15	Rat	Backward chaining	Pierrel and Sherman (1963)
3	Chimpanzee	Forward chaining	Tomonaga and Matsuzawa (2000)
~4	Capuchin monkey		Addessi et al. (2008)
22–25	Otter (*L. canadensis*)	Forward chaining	Bob Bailey, personal communication*
9	Chimpanzee	Forward chaining	Inoue and Matsuzawa (2007)
4	Pigeon	Forward chaining	Straub and Terrace (1981)
3	Pigeon	Forward chaining	Terrace (1986)
5	Long-tailed macaque, rhesus macaque	Forward chaining	Colombo et al. (1993)
8	Rat	Forward chaining†	Fountain and Rowan (1995), Fountain (2008)

Note: Sequences consisted of key presses, lever presses, or tricks such as climbing ladders and opening doors. In some cases, the animal was prompted throughout the sequence; in others, it performed the sequence autonomously. Empty cells signify missing information.

*Record sequence length from yearly competition between animal trainers at Animal Behavior Enterprises, Hot Springs, Arkansas.

†Our interpretation, based on method description.

illustrates an experiment in which pigeons were exposed to an environment containing a light, a response key, and a pedal (Wyckoff 1952, 1969). A white light was an ambiguous predictor of reward: when it was on, pecking a response key resulted in a food reward in 50% of trials. On trials when food was available, pressing the pedal turned the light red. On trials when food was not available, a pedal press turned the light green. The pigeons learned to press the pedal and then pecked only if the light turned red. This outcome can be explained as follows. According to the chaining model, the red light acquired stimulus value because it preceded rewarded pecks. Thus the red light could reinforce pressing the pedal, because pressing was followed by the red light 50% of the time. The green light, however, could not reinforce pecking because it never acquired stimulus value. Could it be, however, that pigeons were actively seeking information about the best course of action, rather than just learning the response through chaining? The key to distinguish chaining from information seeking is to realize that, according to the latter, observing either color is equally valuable. In the words of Bloomfield (1972), "bad news is as much news as good news." However, data show that the observing response is maintained solely by the color that predicts food. For example, Dinsmoor et al. (1972) adopted Wyckoff's design but also included sessions

Group 1:	*Red* + *Green* → Peck red → *(60 s)* → Food
	Red + *Green* → Peck green → *(60 s)* → No food
Group 2:	*Red* + *Green* → Peck red → *Yellow (60 s)* → Food
	Red + *Green* → Peck green → *Blue (60 s)* → Food
Group 3:	*Red* + *Green* → Peck red → *Yellow (30 s)* → *Blue (30 s)* → Food
	Red + *Green* → Peck green → *Blue (30 s)* → *Yellow (30 s)* → No food

FIGURE 3.6. Event sequences in Cronin's (1980) experiment on delayed reward.

in which only the red light (predictive of food) or only the green light (predictive of no food) was operational. Pedal presses were maintained only when they produced the red light but ceased to occur when they produced the green light. Later studies have confirmed this finding (reviewed in Fantino 1977).

Chaining can also account for the results of a remarkable experiment by Cronin (1980), illustrated in figure 3.6. Pigeons could peck either a red or a green light. Pecks to red were rewarded with food and pecks to green were not. Ordinarily, this discrimination would be learned readily, but in this experiment the consequence of pecking red or green became apparent only after 60 s. This proved hard, and pigeons ended up pecking red and green equally often. For another group of pigeons, pecks to red caused the light to turn yellow during the delay, while pecks to green caused the light to turn blue. These pigeons learned to peck red about 95% of the time. The interpretation of this finding with regard to chaining is that yellow acquired stimulus value by virtue of preceding food and was thus able to reinforce pecks to red, while blue did not acquire stimulus value (Spence 1947, Grice 1948). Cronin's most intriguing result was obtained with a third group of pigeons. For these pigeons, pecks to red turned the light yellow initially, then blue. Pecks to green turned the light blue initially, then yellow. Thus the color that immediately preceded food (blue) also followed the *incorrect* stimulus (green), while the color that never preceded food (yellow) followed the correct stimulus (red). Pigeons in this group chose red only about 10% of the time, yet it appears that the few rewards collected in this way accorded sufficient stimulus value to blue to reinforce the incorrect pecks to green.

3.8 Chaining in Nature

Chaining can profoundly modify the behavior of many animals in both laboratory and applied settings. It would be surprising if it did not play a role in natural behavior. At the same time, it is more difficult to study learning in the wild. As we discussed in chapter 1, all but the most trivial environments are likely to provide opportunities for many profitable sequences of actions. However, our discussion in section 3.5 shows that chaining can learn only sequences that are relatively short, ones that have favorable entry or exit patterns, or ones that are already partly known (see chapter 4). To determine whether chaining is a viable learning mechanism in nature would require detailed analysis of natural environments, which we cannot provide for many reasons, including limited time, space, and knowledge of environmental details. Nevertheless, we can consider a few generalized scenarios that demonstrate how chaining may work in nature. We have chosen foraging, hunting, self-control, and tool use because they have been studied extensively and because they pose a range of challenges to learning mechanisms, including the need to give up small, immediate rewards for larger ones in the future.

3.8.1 Foraging

Foraging animals must locate and handle static food sources, like fruit, seeds, and leaves. The chaining model can learn to forage as long as the handling phase is not too demanding, or provided that favorable entry and exit patterns apply. For example, a fruit-eating bird may need to learn a sequence such as the one in equation 3.2. This is feasible given that fruit are encountered often enough while exploring the environment. Moreover, learning will be easier if open fruit are sometimes encountered (a favorable entry pattern) and if mistakes are not catastrophic, such as having the ability to peck again following a misdirected first peck (a favorable exit pattern). However, foraging sequences are often more complex than the one in equation 3.2. For example, black rats can learn to feed on pine seeds that are accessible only after stripping the scales off the pine cone (Terkel 1996). The behavior sequence acquired by the rats consists of firmly grasping a closed pine cone to enable bending and detaching of a cone scale with their teeth, then eating the exposed seed:

	Stimulus:		Behavior:	
Step 1:	*Pine cone (far)*	→	Approach	→
Step 2:	*Pine cone (near)*	→	Grasp	→
Step 3:	*Pine cone (grasped)*	→	Bite scale	→
Step 4:	*Pine cone (bitten)*	→	Detach scale	→
Step 5:	*Seed exposed*	→	Eat	→
Step 6:	*Reward*			

This sequence is repeated until all seeds have been extracted. Learning this foraging technique is challenging because encountering a pine cone does not give any immediate hint that it is a food source. Unless the seeds are discovered early on, the actions of approaching and handling pine cones go unrewarded, and eventually the rat ceases to engage with them. Nevertheless, wild black rats learn to be proficient pine cone foragers.

To elucidate how such learning occurs, Aisner and Terkel (1992) experimentally manipulated the entry pattern to the sequence. Their experimental conditions included young pups and adults presented with cones that had been stripped of a varying number of scales. When provided with closed pine cones, no rat (out of 127 across three experiments) learned the sequence within a three-month period. However, rats that were given access to partially opened pine cones with exposed seeds were often successful, and eventually could open closed cones on their own. Only young rats raised by proficient mothers (including foster mothers, to control for genetic effects) consistently learned the whole sequence. These young rats would gather around the foraging mother and could sometimes successfully steal and eat a seed. Older pups sometimes stole partially opened cones from the mother, carrying them to safety and trying on their own to extract the seeds. This pattern of acquisition, starting from eating seeds, then handling partially opened cones, and finally being able to initiate the opening process, is consistent with an underlying learning process of chaining, facilitated by a favorable entry pattern that allows young rats in their natural environment to practice all steps of the sequence. (These experiments are often discussed in the context of social learning, which we cover in chapter 6.)

3.8.2 Hunting

Another common scenario is that of a predator hunting for prey. A simplified behavior sequence for this scenario is outlined below:

	Stimulus:		Behavior:	
Step 1:	*Prey spotted*	→	Launch attack	→
Step 2:	*Prey flees*	→	Chase	→
Step 3:	*Prey near*	→	Catch	→
Step 4:	*Caught prey*	→	Kill	→
Step 5:	*Prey dead*	→	Eat	→
Step 6:	*Reward*			

(Each behavior could be further decomposed into smaller units.) Learning to hunt is challenging because the interaction with the prey generates unfavorable entry and exit patterns. The sequence is almost exclusively entered from the first step because prey seldom wait to be captured (unfavorable entry pattern). Moreover, mistakes typically result in the prey escaping (unfavorable exit pattern). Various circumstances, however, ameliorate these difficulties and make the sequence learnable. Many predators have genetic predispositions for chasing moving objects (see chapter 4), which solves step 2 in the sequence. For example, inexperienced polecats spontaneously chase escaping prey, but will not hunt stationary prey (Eibl Eibesfeldt 1975). Experienced polecats, on the other hand, attack stationary prey as well.

Fox's (1969) study of canids elucidates how this transition may occur. Like polecats, his hand-raised gray fox and wolf pups spontaneously followed young rats. They also pawed at, sniffed, bit, and eventually killed the rats, suggesting predispositions that aid in establishing sequence steps 3 and 4. At first, however, the pups did not eat the rats they killed, even though they would eat rat meat provided by the caretaker. When they were approximately one month old, however, one pup of each species happened to taste the blood of a kill, and eventually started to eat from it, thus performing sequence step 5. Thereafter, they regularly ate their kills.

According to the chaining model, the experience of eating the killed rat would contribute to establishing sequence step 4 and would also endow the preceding stimulus (sight of prey) with value. Because the prey is present throughout the sequence, the acquisition of stimulus value can reward correct actions and help master all sequence steps. In nature, pups become similarly familiar with prey via their parents. Thus, with appropriate genetic predispositions and a favorable environment, predators can acquire a complex behavior sequence through chaining.

3.8.3 Self-Control

In studies of self-control, researchers have investigated the ability to inhibit a behavior that appears immediately beneficial in order to pursue a larger benefit later (Logue 1988, MacLean et al. 2014). For example, animals may avoid responding immediately to food if this enables them to obtain more food (Grosch and Neuringer 1981, Logue 1988, Evans and Westergaard 2006, Raby and Clayton 2009, MacLean et al. 2014). Work in behavioral ecology shows that this kind of self-control is common in nature (fish: Werner and Hall 1974; crustaceans: Elner and Hughes 1978; birds: Krebs et al. 1977, Goss-Custard 1977). In addition, experienced predators, such as adult cheetahs (*Acinonyx jubatus*), can inhibit attacking prey in order to increase hunting success, whereas inexperienced cheetahs often attack when prey is vigilant or too far (Schaller 1968, Caro 1994, Hilborn et al. 2012).

Self-control is often considered an ability beyond associative learning, but this is not true in general. Chaining can incorporate self-control because it can learn to value any behavioral option, including waiting or ignoring food. If these options are better in the long run, chaining learns them as part of

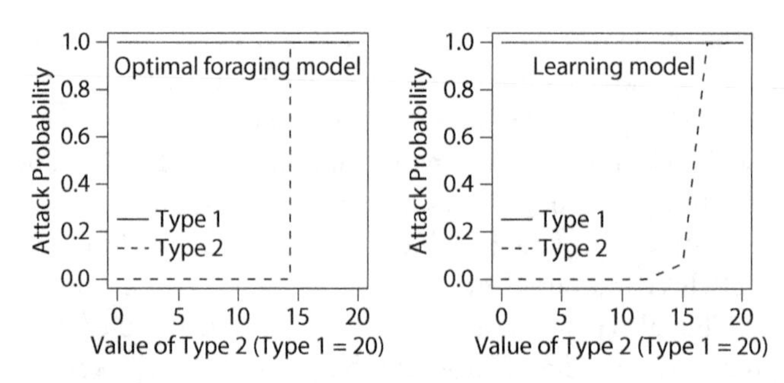

FIGURE 3.7. Example of learned self-control in prey selection. A predator can encounter two types of prey: large (type 1) and small (type 2). Left: According to optimal foraging theory, large prey should always be attacked, but small prey should only be attacked if its value is high enough (Stephens and Krebs 1986, section 2.2). Otherwise, the time spent handling small prey is better spent continuing to look for large prey. Right: Chaining can learn to closely approximate the optimal strategy, forgoing the immediate reward offered by small prey for the opportunity to continue looking for large prey when this is warranted. See section 1.5.5 for simulation details.

its ability to optimize behavioral sequences rather than just single behaviors (Enquist et al. 2016, Lind 2018). For example, figure 3.7 considers a classic scenario from behavioral ecology. A predator can encounter two types of prey: large and small. Large prey should always be attacked; but whether small prey is worth attacking is a more difficult decision, because spending time on small prey implies finding fewer large prey. According to optimal foraging theory, this is a complex decision involving the value of small and large prey, the time required to capture and consume them, and their abundance (Stephens and Krebs 1986). Figure 3.7 shows that chaining approximates the optimal behavior closely, learning to ignore small prey when its value is below the threshold identified by optimal foraging theory. In this example, chaining learns to inhibit an action with immediate value that is detrimental in the long run, taking into account such factors as the abundance of the two types of prey, their value, and their handling time.

3.8.4 *Tool Use*

Here we consider whether chaining is a plausible mechanism to learn tool use. Broadly defined, tool use occurs in a variety of distantly related species (Bentley-Condit and Smith 2010). For example, red-tailed hawks (*Buteo jamaicensis*) bash captured snakes against a rock (Ellis and Brunson 1993); *Ammophila* wasps use a pebble to flatten the soil around their burrow's entrance (Brockmann 1985); chimpanzees crack nuts with stone hammers

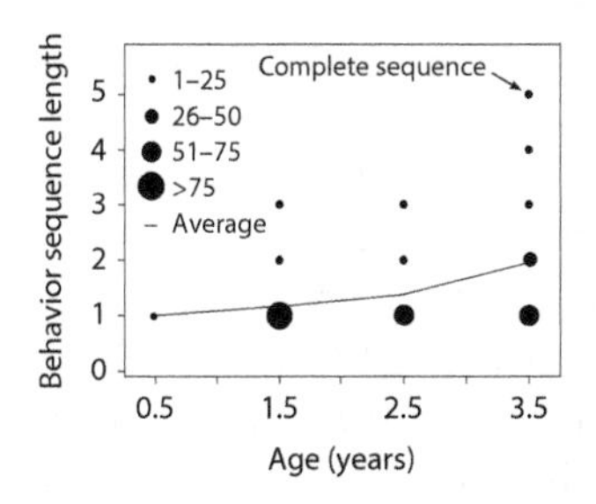

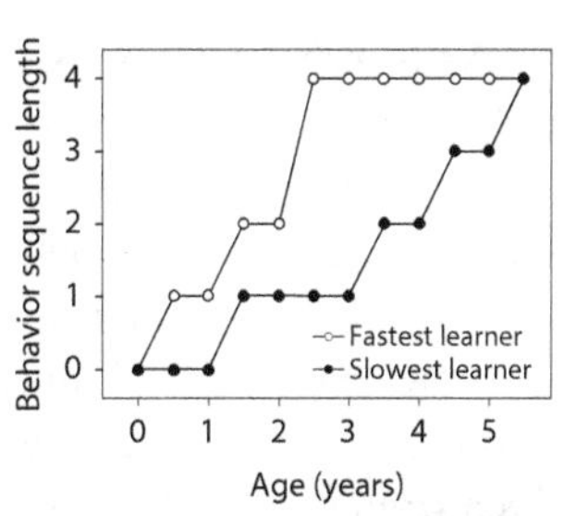

FIGURE 3.8. Development of tool use in wild chimpanzees. Left: Development of nut cracking with stone tools in three chimpanzees (Inoue-Nakamura and Matsuzawa 1997, table 7). Functional tool use requires a sequence of five different behaviors in the correct order (take a nut, place it on the stone anvil, hold a stone hammer, hit the nut with the hammer, eat the nut). Complete sequences were first observed around 3.5 years of age, but only at a very low frequency. Most behavioral sequences included fewer than two behaviors on average. Right: Development of termite fishing in the slowest and fastest of eight chimpanzees (Lonsdorf 2005, table 2). Termite fishing was described as a sequence of four behaviors: find a termite hole, make a tool from a twig, insert the tool into a hole, and extract the tool.

on a rock anvil (Inoue-Nakamura and Matsuzawa 1997). Some cases of tool use involve little learning (such as in wasps), while others include some of the longest sequences that animals learn in nature. We start with one of the most remarkable examples, nut cracking in chimpanzees, and then come back to the general issue of how tool use may be learned.

Nut cracking has been observed in some chimpanzee groups (Sugiyama and Koman 1979) and can be described as a sequence of five behaviors: picking up a nut, putting it on a stone (anvil), taking a second stone (hammer), hitting the nut with the second stone, and finally eating the nut (Inoue-Nakamura and Matsuzawa 1997):

	Stimulus:		Behavior:	
Step 1:	*Nut*	→	Pick up	→
Step 2:	*Nut in hand*	→	Put nut on anvil	→
Step 3:	*Nut on anvil*	→	Take stone hammer	→
Step 4:	*Nut on anvil+ stone in hand*	→	Hit nut	→
Step 5:	*Open nut*	→	Eat	→
Step 6:	*Reward*			

Chimpanzees learn this sequence with great difficulty. It is fully mastered only at around five years of age, and about 20% of individuals never learn it (figure 3.8). Similar tool-based techniques are learned equally slowly, such as using sticks to fish for termites or cupping leaves to drink (Nishida and Hiraiwa 1982, Lonsdorf 2005, Sousa et al. 2009). Nevertheless, upon reflection these learning feats are quite remarkable. For example, chimpanzees have been observed to use about 40 behaviors when interacting with stones and nuts. These include behaviors useful for nut cracking, but also others like mouthing stones, biting closed nuts, or banging them on the ground. With these behaviors, chimpanzees could form $40^5 \simeq 100$ million sequences of five actions. Even trying out 1000 randomly picked sequences per day, it would take almost 300 years to stumble upon the correct sequence. Thus chimpanzees learn much more efficiently than if their behaviors were solely based on random trial and error. In this section, we show that chimpanzees may learn nut cracking purely through chaining. This conclusion agrees with findings that chimpanzees can learn nut cracking and other tool use behavior without seeing others demonstrate the behavior (Reindl et al. 2018, Bandini and Harrison 2020). Nevertheless, the social environment can facilitate learning by providing favorable entry patterns to tool use, as we discuss below and in chapter 6.

Similar to black rat pups (section 3.8.1), young chimpanzees sometimes obtain nuts cracked by their mothers or other individuals. These experiences establish step 5 of the sequence and endow the sight of nuts with stimulus value. Indeed, six-month-old chimpanzees already pick up and manipulate nuts. Until about 2.5 years of age, picking up nuts and eating them are the only behaviors in the nut-cracking sequence that the young perform reliably. During this time, however, we speculate that stones (used as both hammers and anvils) also acquire stimulus value by virtue of their spatial and temporal proximity to the act of eating nuts. The young, in fact, manipulate stones more frequently with age. At this point, a number of plausible effects of stimulus value facilitate the discovery of the nut-cracking technique. For example, placing nuts on a stone (anvil) is reinforced because it creates a stimulus, *Nut on anvil*, that is itself composed of valued stimuli. Moreover, hitting a stone anvil with another stone (hammer) is reinforced because it produces a sound that has predicted many times the availability of open nuts (see marking in section 3.6). The young chimpanzee should thus find it rewarding to manipulate nuts and stones and to hit stones together. It seems plausible that she will eventually discover how to crack nuts by chance. Indeed, computer simulations indicate that chaining can learn the nut-cracking sequence in a realistic 3.5–5 years, assuming about 10 trials per day (Enquist et al. 2016). Figure 3.9 shows a sample simulation, which also demonstrates a feature of chaining that is seldom noted. Namely, little learning is apparent to an external observer until right around the time when most actions are learned, at which point the probability to perform a correct sequence rises quickly from almost 0 to

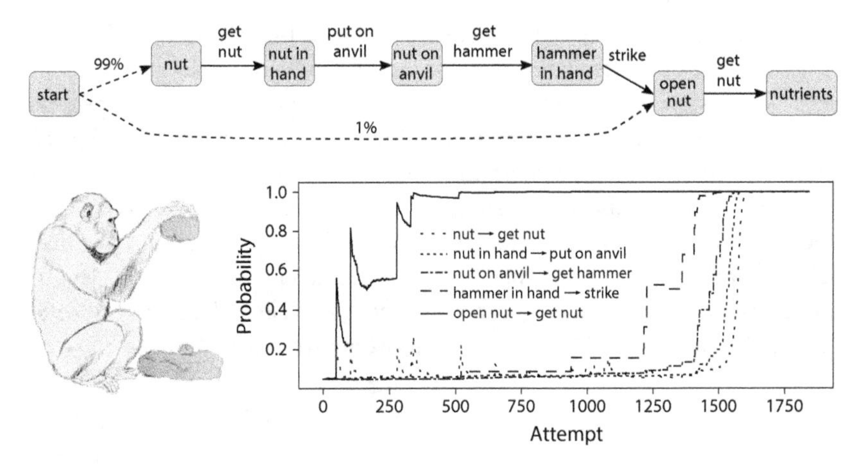

FIGURE 3.9. Learning nut cracking through chaining. Top: Schematic depiction of the nut-cracking task. Bottom left: Artist's depiction of the striking step (Snyder 2018). Bottom right: Probability of performing each action at the correct step in the sequence. See section 1.5.5 for simulation details.

almost 1. In other words, while learning through chaining is stepwise and gradual, it can mimic in overt behavior the suddenness of insight learning (see section 7.7.1). Studies in other primates also suggest that tool use may be learned through chaining. For example, capuchin monkeys manipulate objects extensively from an early age, yet they take years to learn functional tool use (Visalberghi 1987, Visalberghi and Limongelli 1994). In one study, only 2 out of 9 monkeys used tools before age three (De Resende et al. 2008); in another, only 2 out of 42 monkeys acquired tool use behavior after extensive training (Visalberghi 1987).

In summary, our analysis suggests that access to open nuts is crucial to learning nut cracking, in the same way that black rats learn how to open pine cones only if they can feed on partially opened ones (section 3.8.1). Before the first chimpanzee invented nut cracking, access to open nuts was exceptionally rare, which may explain why nut cracking occurs only in some chimpanzee groups. Generalizing this argument, we can see why learned tool use is rare. Using a tool necessarily adds behaviors at the beginning of a sequence, including locating, fashioning, and using the tool, and lengthening a sequence increases learning times dramatically (section 3.5). Thus it seems that tool use can appear only under favorable circumstances. Primates, for example, have prolonged juvenile periods during which the young can explore their environment, and they live in social groups supporting favorable entry patterns for tool use sequences, such as availability of open nuts and suitable stones for nut cracking. Visalberghi (1990, 149), for example, has described the acquisition of tool use in capuchin monkeys as the relentless exploration of all possible behaviors involving nuts and the tool. In the absence of favorable circumstances, we expect tool use to be supported by strong genetic predispositions, such as seeking out and manipulating appropriate objects, which we discuss in chapter 4. Animals living in human environments are also provided with favorable circumstances for chaining, such as readily available tools, rewards for engaging in tool use, and ample opportunities for practice. These considerations may explain why animals living with humans seem to use tools more readily than wild conspecifics (see Call and Tomasello 1996, Tomasello and Call 2004, for review and alternative explanations).

4

Genetic Guidance of Learning

- Genetic guidance is necessary for productive learning.
- Genetic predispositions can influence choice of behavior, exploratory tendencies, and what an animal learns about.
- Genetic predispositions foster efficient behavioral adaptations, but they also limit what can be learned and can be maladaptive outside of a species' niche.
- With weaker genetic guidance, animals could learn under a wider array of circumstances, but this flexibility would increase learning costs.
- Natural selection often restricts flexibility in order to decrease learning costs.

4.1 Introduction

So far we have discussed one kind of genetic influence on learning: the primary value of stimuli, which determines what animals strive to obtain or avoid. Genes influence learning in other ways, too. For example, rats that experience gastric illness learn to avoid the flavors they tasted, but not the sounds and sights experienced while eating (Garcia and Koelling 1966, Seligman 1970). Genes also influence exploration and decision making. For example, many tool-using species are predisposed to manipulate objects that can serve as tools, and hungry animals are predisposed to enact behaviors conducive to feeding (Ewer 1968, Eibl Eibesfeldt 1975, Kenward et al. 2005, Koops et al. 2015). Genetic predispositions are of special interest to us because they play a double role in the evolution of intelligence. On one hand, they can make learning and behavior efficient in the animal's ecological niche (Hinde and Stevenson-Hinde 1973, Revusky 1984, Timberlake 1994, Domjan 2012). On the other hand, non-predisposed behavior becomes more difficult to learn, making the animal less flexible. The capacity to learn under many different circumstances is a hallmark of intelligence, and in this chapter we discuss the trade-off between learning efficiency and behavioral flexibility. Concretely, we consider how genes influence memory updates, decision making, and exploration. We conclude that natural selection typically acts to restrict behavioral flexibility in order to decrease learning costs.

4.2 Memory Changes

Genetic predispositions can determine which aspects of an experience are learned. For example, Foree and LoLordo (1973) trained two groups of pigeons to press a treadle upon simultaneous presentation of a light and a sound. For one group, treadle pressing yielded food; for the other, it postponed a small electrical shock. After learning to respond, the pigeons were presented with the light and the sound separately. Pigeons that had pressed to earn food responded much more to the light than to the sound, reflecting that pigeons rely on sight to identify food, such as seeds and bugs. Pigeons that had pressed to avoid shock, on the other hand, responded more to the sound. Thus the two groups learned different things despite performing the same response to the same stimuli. Similar results have been obtained many times, showing that genetic predispositions influence how easily an association is learned (Garcia and Koelling 1966, Seligman 1970, LoLordo 1979, Cook and Mineka 1990). In another example, Shettleworth (1975) showed that food can reinforce actions relevant to foraging in hamsters, such as scrabbling and locomotion, but not unrelated actions, such as grooming.

To include these genetic predispositions in the chaining model, we can let stimuli and responses influence learning rates (Rescorla and Wagner 1972). Namely, we can replace the single parameter α_v in equation 3.12 with a function $\alpha_v(S, B, S')$ that can yield different values depending on what stimuli and behavior occur. When learning rates differ for two stimuli that are simultaneously present, the stimulus with the higher learning rate accrues a stronger association with the response. For example, we can reproduce the results of Foree and LoLordo (1973) by assuming that under food reinforcement the *Light* → Press association is learned much faster than the *Sound* → Press association, and vice versa when the reinforcer is shock postponement. Formally, we have

$$\alpha_v(\textit{Light}, \text{Press}, \textit{Food}) \gg \alpha_v(\textit{Sound}, \text{Press}, \textit{Food})$$

$$\alpha_v(\textit{Light}, \text{Press}, \textit{No shock}) \ll \alpha_v(\textit{Sound}, \text{Press}, \textit{No shock})$$

It is also possible to vary the learning rate for stimulus values (α_w in equation 3.11) according to S and S', thus influencing which stimuli can become conditioned reinforcers. For example, it seems plausible that a stimulus could become a conditioned reinforcer for a particular response only if it can itself be associated with that response. We are not aware, however, of any experiments that test this possibility.

4.3 Decision Making

While most studies on genetic predispositions in learning have focused on memory changes (Hinde and Stevenson-Hinde 1973, Roper 1983), genes can also influence decision making. These effects are impactful because learning requires behavioral exploration. Genetic predispositions in decision making can greatly decrease exploration costs by focusing the animal on behaviors that are more likely to be useful. For example, a hungry rat will be more likely to find food if it primarily uses behaviors such as moving around, scratching the ground, or poking with its nose, as opposed to behaviors such as grooming or sleeping, which have a negligible chance of leading to food. Recall that learning a sequence of l behaviors, each chosen out of m possible ones, requires m^l attempts (equation 1.1). If the animal knows that only a fraction a of all behaviors are included in the sequence, however, the learning time is cut by a factor of a^l. For example, with $l = 3$ and $a = 1/2$ we would have to explore only a fraction $1/2^3 = 12.5\%$ of the total number of possible sequences. This knowledge can be encoded genetically through what we call *context-dependent* behavioral repertoires, meaning that only a subset of behaviors is likely to occur in particular contexts, such as foraging, mating, or fleeing (Timberlake 1983, Revusky 1984, Timberlake 1994, Timberlake and Silva 1995, Timberlake 2001). For example, sea otters use rocks to crack shells, and the frequency of rock manipulation increases when the otter is hungry (Pellis 1991, Allison et al. 2020).

Context dependence is also seen in learning (Bouton 2016, chapter 10). For example, pigeons that learn to peck a key to obtain water approach the key delicately, with open eyes and semiclosed beak, as they would do to drink. Pigeons that learn to peck for food, however, strike at the key vigorously with open beak and closed eyes, as if pecking at grain (Moore 1973, Jenkins and Moore 1973). Similarly, Farris (1967) observed that male quail would display courting behaviors when approaching a key they had learned to press to access a mate (see also Domjan 1992).

A behavioral context can be further divided into subcontexts with even smaller repertoires. For example, foraging involves a search phase during which locomotion and inspection of objects are appropriate, followed by a separate consumption phase, during which eating behaviors are required. Consistent with this analysis, rodents provided with access to a running wheel and food may run for a considerable time before eating, even though the food is always available (Levitsky and Collier 1968). At any given time, the active behavioral context can depend on many factors, such as the presence of specific stimuli, previously performed behaviors, and the animal's motivational state (McFarland and Houston 1981, Timberlake 2001). In general, the stronger the predisposition for the correct action, the faster a sequence is learned. The downside is that strong predispositions make it difficult for the animal to learn what it is *not* predisposed to do. For example, Breland and Breland (1961) trained pigs to deposit tokens into a "piggy bank," using food as a reinforcer. Training was initially successful, but after a while the pigs started to handle the tokens as food and became reluctant to let go of them. We can interpret this behavior as the token becoming associated with food and thus eliciting behavior that is normally directed toward food. Especially in artificial environments, the tendency to behave in a certain way can be so strong as to result in nonfunctional behavior (see also Timberlake et al. 1982).

Context-dependent behavior can be introduced into models of associative learning by developing the decision-making process. In the chaining model, choice of behavior depends partly on stimulus-response values and partly on the β parameter in the decision-making function (equation 3.8). We can replace the single parameter β with a function that favors different responses depending on the current stimulus, current motivation, and previous responses. Thus the β value can be small or zero for presumably unproductive actions and high for productive ones (Enquist et al. 2016). Because β multiplies stimulus-response values, this approach effectively gives more value to more appropriate responses. For example, the reluctance to relinquish items associated with food, observed by Breland and Breland (1961) in their pigs, can be reproduced by assuming that, in the context of feeding, β is higher for the response of holding on to objects, rather than for letting go of them (Enquist et al. 2016).

4.4 Exploratory Tendency and Behavioral Repertoires

Genes also regulate exploration of the environment, which is an important part of decision making (section 3.2) and can have a profound effect on learning. Empirical data show that exploratory behavior varies significantly among animals (figure 4.1). Reptiles, for example, show little interest in novel objects. Carnivores and primates explore novel objects extensively, but carnivores employ stereotyped motor patterns, such as chewing and pawing, whereas some primates show considerable diversity in motor patterns, including individual differences (Glickman and Sroges 1966). Overall, exploratory tendency appears to correlate with behavioral flexibility: while most species rely on a small repertoire of behavioral patterns that are programmed genetically (Lorenz 1941, Hinde 1970, Hogan 2001), more exploratory species

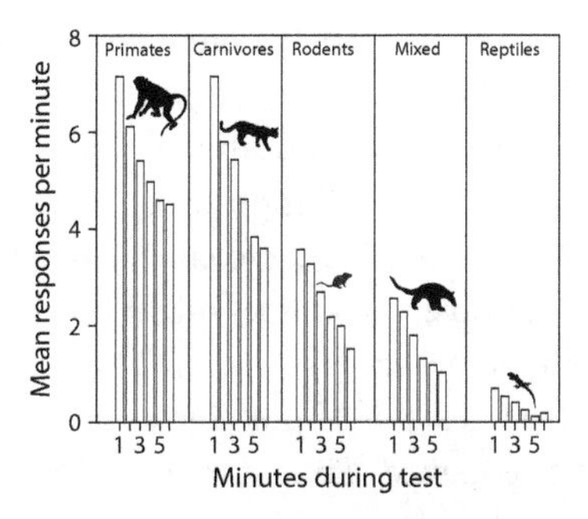

FIGURE 4.1. Exploration of novel objects by zoo animals (redrawn from Glickman and Sroges 1966). The "mixed" group includes marsupials, insectivores, anteaters, and armadillos. Large variations existed also within taxa. In primates, for example, chimpanzees and Allen's swamp monkeys (*Allenopithecus nigroviridis*) explored seven to nine times more than patas monkeys (*Erythrocebus patas*) and colobus monkeys. Overall, young explored more than mature animals, and no sex differences were found.

Table 4.1. Number of behavior types in mammals

Animal order (groups with data)	Behavior types (species)
Primates (lorises, monkeys, apes)	44–184 (7)
Cetacea (dolphins)	123 (1)
Chiroptera (bats and fruitbats)	93 (1)
Carnivora (ferrets and cats)	69–74 (2)
Rodentia (rats, squirrels, maras, and cricetids)	24–55 (12)
Insectivora (shrews)	54 (1)
Artiodactyla (bovids and deer)	22–32 (2)

Source: Changizi (2003).

tend to have larger and more flexible repertoires (table 4.1). Primates, elephants, and octopuses are examples of animals that have both exceptional manipulation abilities and large behavioral repertoires (Parker 1974, 1978, Chevalier-Skolnikoff 1989, Chevalier-Skolnikoff and Liska 1993, Christel 1993, Mather and Anderson 1999, Kuba et al. 2006). As recalled above, the chaining model can account for different exploratory tendencies by assuming that genes code for different β values in different species. The connection between exploration and behavioral flexibility can be clarified as follows.

A species' behavioral repertoire is subject to two conflicting selection pressures. A larger repertoire can benefit the animal by enabling it to act on its environment in more ways, potentially allowing access to more resources. The downside of a large behavioral repertoire is that more environmental exploration requires more time. A simple model can illustrate this cost-benefit trade-off (Ghirlanda et al. 2013). Suppose the animal is trying to learn a rewarding sequence of l actions, in order to collect as much reward as possible in a fixed time T. For example, the animal might be a fledgling bird trying to eat as much as possible before its first winter. The actions that make up the sequence are chosen from a set of M possible actions, but the animal has no information about the sequence. Thus it must try out all sequences until it finds the rewarding one. At this point, it continues to perform the rewarding sequence for the time that remains.

The animal's repertoire might include all possible actions, but this can be suboptimal as it entails a large learning time. In fact, there are M^l possible sequences of length l, and this number is also the expected time to find a rewarding sequence when trying out sequences randomly (section 1.3.1). It turns out that it is typically better to reduce the learning time by using a smaller repertoire $m < M$, even if this introduces the risk of not being able to perform the target sequence at all. This can be shown by calculating the optimal repertoire size, which is the one that maximizes the expected number of times that the rewarding sequence is performed. The latter can be written as follows:

$$E(\#\text{ rewards}) = \Pr(\text{reward is accessible}) \times [\#\text{ available attempts} - E(\#\text{ learning attempts})]$$

where E denotes expected value. The first term is the probability that the rewarded sequence is within the scope of the animal's repertoire. This number is m^l/M^l, as the number of sequences that can be performed with a repertoire of size m is m^l, out of M^l possible sequences. The number of available attempts is the number of sequences of length l that can be performed in time T. If we assume for simplicity that each action takes one unit of time, this number is the integer part of T/l, which we write as $\lfloor T/l \rfloor$. The expected number of learning attempts is m^l. Putting everything together, we find

$$E(\#\text{ rewards}) = \frac{m^l}{M^l} \times \left(\left\lfloor \frac{T}{l} \right\rfloor - m^l + 1 \right) \tag{4.1}$$

where the $+1$ term takes into account that the rewarding sequence is also performed during the last learning attempt. As m grows from 1 to M, the first term in equation 4.1 increases, corresponding to a higher chance that the rewarded sequence can be performed, while the second term decreases, because the time to locate the sequence increases. The trade-off between these two factors

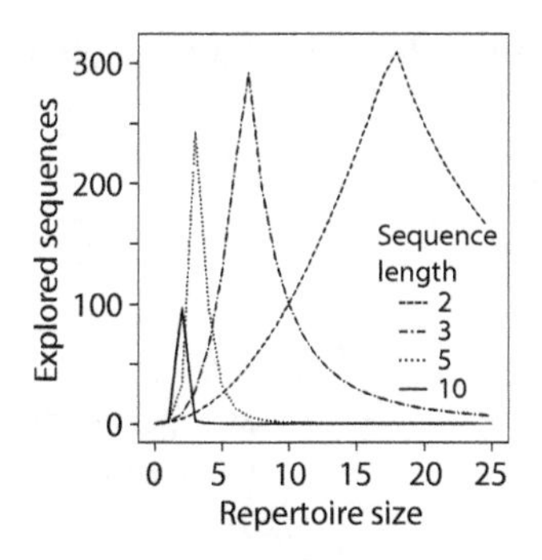

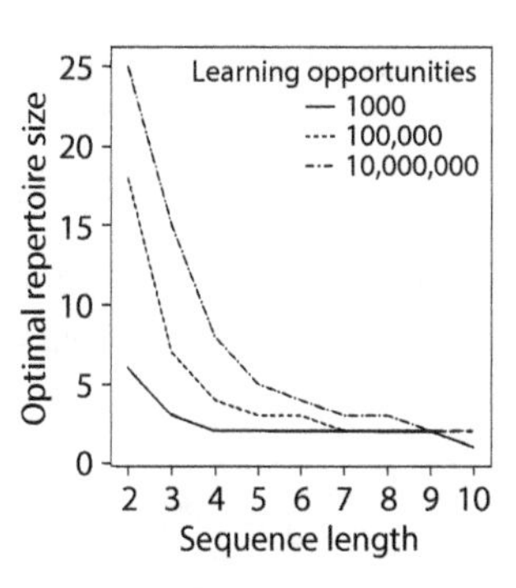

FIGURE 4.2. Left: Number of behavioral sequences of a given length that can be explored with different repertoire sizes, given that $T = 10^5$ actions can be explored in a lifetime. Right: The repertoire size that maximizes the number of rewards collected, as a function of sequence length, for three different numbers of learning opportunities, based on equation 4.1.

is depicted in figure 4.2 for some values of T and l. The optimum decreases with l and increases with T, and it is typically quite small aside from very short sequences. Thus, even if the genes can in principle allow large behavioral repertoires (more behavioral flexibility), it seems that natural selection would not favor this strategy. Rather, it would be better to be able to recognize contexts and stimulus situations and associate to each one a small repertoire specialized for that context. As recalled in the previous section, this seems to be the case in most species.

The model can also be considered from the point of view of sequence length. Given an amount of time and a behavioral repertoire, should an animal try to learn short or long sequences? Long sequences can be more rewarding (compare cracking fat-rich nuts to eating readily available plant matter), but they are exponentially more difficult to learn. We expect that in most environments the safest strategy is to focus on learning short sequences. This strategy results in behavior of limited complexity.

In conclusion, flexible behavior is expected only in animals that can invest significant time in learning. For example, Heldstab et al. (2020) compared the motor development of 36 primate species, and found that species that are more apt at manipulation require more time to develop adult-level skills. They concluded that complex skills, such as foraging with tools, require a long juvenile period during which the offspring is protected while it explores its environment and benefits from social interactions. Many other studies have reported compatible findings in the context of foraging and hunting skills (Ewer 1968, Fox 1969, Inoue-Nakamura and Matsuzawa 1997, Kenward et al.

Table 4.2. Genetic predispositions in the A-learning model of chaining

Parameter	May depend on	Influences	Enables genes to control
β, response bias	S, B	Selection of behavior	Amount of exploration Response bias (different βs for different B) Context dependence (different βs for different S, B)
u, primary value	S'	Maximum values of v and w	What responses can be learned, and how fast What to strive for (what is a reinforcer)
α_v, learning rate	S, B, S'	Rate of change of v	What responses can be learned ($\alpha_v > 0$), and how fast Context dependence (different α_vs for different S, B, S')
α_w, learning rate	S, B, S'	Rate of change of w	What S can acquire value ($\alpha_w > 0$), and how fast Context dependence (different α_ws for different S, B, S')

Note: The table describes how genes may influence the parameters in equations 3.8, 3.11, and 3.12, when these parameters can depend on stimuli S and S' and on behavior B. This notation refers to the sequence $S \to B \to S'$ in which stimulus S is experienced, then behavior B is performed, and then stimulus S' is experienced.

2006). A prolonged juvenile period with protected exploration appears characteristic of all behaviorally flexible animals (Poirier and Smith 1974, Cutler 1976, Scarr-Salapatek 1976, Lovejoy 1981, Harvey and Clutton-Brock 1985, Ghirlanda et al. 2013a).

4.5 Adapted Flexibility

Typically, genetic influences promote adaptive behavior by focusing the learning process on relevant stimuli and behaviors, thereby decreasing learning times. As mentioned earlier, for example, foraging pigeons learn more quickly about visual compared to auditory stimuli, because they identify food by sight (section 4.2). In general, we suggest that natural selection favors genetic predispositions that balance behavioral flexibility and learning time. In other words, we expect flexibility to evolve only if it results in benefits that outweigh the cost of increased learning times. We also expect predispositions to reflect a species' ecological niche. For example, exploratory tendencies may be favored in environments with few predators, while predator recognition is favored when predators are common (Jolly et al. 2018, Lapiedra et al. 2018). Furthermore, we speculate that much cross-species variation in behavior derives from differences in genetic predispositions, morphology, and behavioral repertoires, rather than from differences in learning mechanisms (Enquist et al. 2016). If this is true, a general learning mechanism, such as the chaining model, might account for species-typical behavior by including species-specific values of parameters, as summarized in table 4.2. To illustrate the wide-ranging consequences of different predispositions, we can compare two species that do not use tools in the wild, rooks and bonobos, with their close tool-using relatives, New Caledonian crows and chimpanzees. We could conclude that rooks and bonobos do not use tools because they lack the necessary learning ability, but these species can readily be trained to use tools in captivity (Toth et al. 1993, Bird and Emery 2009). Rather, rooks and bonobos lack a predisposition to manipulate objects that can serve as tools, whereas New Caledonian crows and chimpanzees show a keen interest in object manipulation from a very young age, which promotes learning of tool use (Kenward et al. 2006, Furuichi et al. 2015, Koops et al. 2015, Rutz et al. 2016). The hypothesis that tool use is developed through genetic predispositions rather than through specific learning mechanisms is also supported by studies of other species (Thouless et al. 1989, Parker and Poti 1990, Westergaard et al. 1998, Tebbich et al. 2001, Kenward et al. 2005, Tebbich et al. 2010).

5

Sources of Information

- Animals gather and store information to fulfill diverse needs.
- Animals use information from stimuli, memories, and internal factors, like motivational states.
- General-purpose working memory in animals is limited: it encodes sequential structure (such as the order of events) poorly, and it rarely lasts for more than a few minutes.
- However, many species possess memory specializations to store certain biologically relevant information, such as food locations.
- Specialized memories can make an associative learning mechanism more capable without changing how it learns.

5.1 Introduction

Decision making, whether in animals, humans, or machines, combines information about the present situation with information stored in memory. In this chapter, we survey what information animals memorize and use. In section 5.2, we recall that animals are often very skilled in using information from sense organs as well as internal stimuli. In section 5.3, we discuss how animals extract from experience statistical information about the world. The next sections consider memories for specific events—both single events and sequences of events (section 5.4)—and specialized memory systems that only deal with certain kinds of information (section 5.5).

Memory is an important topic because the impressive memory abilities of animals indicate a sophisticated capacity to represent the external environment, which has been deemed incompatible with associative accounts of animal intelligence (Terrace and Metcalfe 2005). However, while sophisticated representations can be used for thinking, they can also be used by associative learning, as we show in section 5.6.

Before delving into the diversity of animal memories, we recall that the ultimate reason for this diversity is that remembering everything that ever happens is a losing game. Not only would it require an implausibly large memory, but it would also make it harder to attribute current events to past causes, as a multitude of potential causes would have to be considered. Having to consider large numbers of experiences would also make decisions more time consuming. Thus, as observed in section 1.3, learning and memory must be selective, and tailored to each species' needs (Shettleworth 1972, Hinde and Stevenson-Hinde 1973, Bouton 2016, Lind et al. 2021).

5.2 Stimulus Recognition and Discrimination

In countless studies, animals have been found capable of recognizing and discriminating stimuli based on features such as intensity, color, sound frequency, size, shape, and even abstract relations such as identity (Ghirlanda and Enquist 2003, Pearce 2008). Thus animals are considerably more flexible in learning responses to stimuli than in learning motor patterns (chapter 4), although it may take time and genetic predispositions may still be important. It seems fair to say that research on stimulus recognition has failed to reveal any major differences between animals and humans, other than innate recognition being more important in animals (Tinbergen 1951, Eibl Eibesfeldt 1975).

Animals can recognize objects under many conditions, such as against different backgrounds, from different points of view, and under different lighting (object constancy; see, for example, Vallortigara 2004). Responses to a stimulus generalize appropriately to similar stimuli (Ghirlanda and Enquist 2003). For instance, a pigeon trained to peck on a green key will also peck similarly colored keys (Guttman 1959, Hanson 1959). When needed, animals pay attention to small details. For example, yellowfin tuna can detect small differences in magnetic field intensity (Walker 1984), and pigeons can detect visual differences that are not easily apparent to human observers (Hodos and Bonbright 1972). However, animals can also learn to treat different stimuli as members of the same category. After learning to discriminate between 10 Picasso and 10 Monet paintings, pigeons could discriminate between Picasso and Monet paintings they had never seen before (Watanabe et al. 1995). Animals ranging from insects to birds and mammals can also learn to respond according to abstract categories, such as discriminating between similar and dissimilar objects or symmetrical and asymmetrical ones (Robinson 1955, Delius and Habers 1978, Delius and Nowak 1982, Delius 1994, Giurfa et al. 1996, Giurfa et al. 2001). These recognition and discrimination abilities are often coupled with large memory capacity. Pigeons can remember hundreds of pictures for years, and baboons thousands (Vaughan and Greene 1984, Fagot and Cook 2006). The dog Chaser could respond correctly to the English name of more than 1000 objects (Pilley and Reid 2011; see also Griebel and Oller 2012). Finally, animals can integrate sensory information from different sources. Dolphins, for example, integrate visual and acoustic information when recognizing objects (Harley et al. 1996), and internal senses (such as circadian clocks and motivational states) can also be integrated with information from external stimuli (Biebach et al. 1989, Hogan and Van Boxel 1993, Pahl et al. 2007).

The ability to recognize and distinguish stimuli has a strong genetic basis grounded in architectural features of the nervous system, such as the pathways that bring retinal signals to visual areas and the organization of these areas (Ewert 1980, Hubel 1988, Gegenfurtner and Sharpe 1999). At the same time, perception is often tuned through experience (Goldstone 1998, Espinosa and Stryker 2012, Dosher and Lu 2017), which can be important to understand some learning effects (see, e.g., section 7.6). In particular, even non-reinforced experiences with particular stimuli can increase an animal's ability to learn about them, a phenomenon called *perceptual learning* (Gibson 1969, Hall 1991, Goldstone 1998). For example, rats learn more easily to discriminate between two visual shapes if the shapes are familiar to them (Gibson and Walk 1956). Perceptual learning is thought to reflect changes in neural processing, such that the representations of stimuli become more precise (Dosher and Lu 2017), which can facilitate both discriminating between stimuli and recognizing similarities (McLaren and Mackintosh 2000, 2002). Conversely, lack of experience with particular perceptual features often leads to deficits (Espinosa and Stryker 2012).

5.3 Statistical Information about the World

Associative learning is the perfect example of how using summary information, rather than the whole history of events, can lead to efficient learning and decision making. In associative learning, experiences are remembered in the form of associations, rather than recorded in every detail. For example, the decision-making equation we considered in chaining (equation 3.8) selects behaviors based only on their current stimulus-response value v (associative strength). In addition, the learning equations for stimulus values (w, equation 3.11) and for stimulus-response values (equation 3.12) only consider small snippets of experience (a stimulus, a behavior, and the next stimulus), rather than extended histories. These values embody statistical information about the environment because they approximate expected values of stimulus situations (w) and of actions (v), which take into account the probability that one stimulus situation gives rise to another (Enquist et al. 2016). That such a simple system can learn to behave optimally under a wide variety of circumstances (chapter 3) is a testament to how much can be accomplished by only remembering statistical summaries of experiences. Indeed, a century of research on Pavlovian and instrumental learning shows that associative learning theory captures well animals' sensitivity to statistical regularities in their environment (Dickinson 1980). There is some debate, however, as to exactly what information may be remembered. In chaining, the value $v(S \to B)$ is modified based on an experience of the form $S \to B \to S'$ and comes to represent the expected value of S'. Yet, after the experience is over, nothing is remembered of S'. Many scientists are dissatisfied with this view, because it seems to imply that animals would perform B in response to S without any clue of the consequence of the response (Roper 1983, Pearce 2008). However, this implication is not entirely correct, as $v(S \to B)$ informs the animal of the expected value of S' if not of its exact nature. From an information standpoint, this would, for instance, be sufficient to explain why animals do not "accept" a sudden decrease in food value, and instead show frustration and search for the expected food (Enquist et al. 2016; see section 3.8.3 for a similar argument in the context of self-control). Likewise, the stimulus value $w(S)$ estimates the expected value of the following stimulus, S'. Associative learning theorists have also entertained the hypothesis that animals learn stimulus-stimulus associations, which would inform animals of the probability that S is followed by a specific S', as well as causal information about the effects of behaviors, which can be learned from $S \to B \to S'$ observations (Bouton 2016, Pearce 2008). We return to this topic in chapter 7.

5.4 Memory for Specific Events

In this section and the next one, we explore the extent to which animals can remember and use information from specific events, rather than just summary statistics from many events. For example, a fox may remember seeing a rabbit enter its burrow a few minutes ago, in addition to having learned that there is a rabbit in the burrow, say, 10% of the time (figure 5.1). When referring to animals, memories of specific events are sometimes called "episodic-like" memories (Clayton et al. 2007). From a functional perspective, there are both advantages and disadvantages with such memories. While they can provide valuable information, they also introduce costs in learning and decision making (see the introduction to this chapter, and section 1.3). Because of these costs, natural selection may favor specialized abilities to remember certain kinds of events of reliable utility to a species, rather than a general ability to remember any event. For example, caching birds can remember hundreds or thousands of food locations, but in other contexts their memory is comparable to that of noncaching species (Olson et al. 1995, Bednekoff et al. 1997,

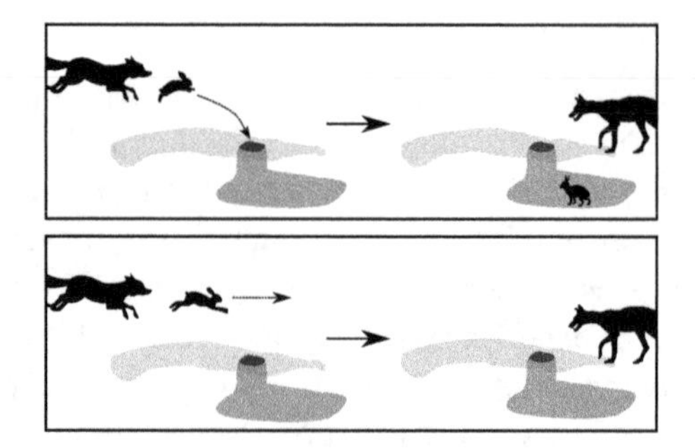

FIGURE 5.1. Top: A fox observes a rabbit escape into its burrow. Bottom: A fox observes a rabbit escape by running away. From just observing the burrow (depicted on the right in both panels), the fox cannot know whether the rabbit is hiding there or not. By remembering a past event—the rabbit entering the burrow or running away—the fox can know whether it is worth waiting in front of the burrow.

Clayton and Dickinson 1999b). In this section, we consider animals' abilities to remember generic events, while in the next section we discuss memory abilities that apply only to specific events. For our discussion, it is also important to distinguish between memories of single events and memories of event *sequences*, the latter being a prominent feature of human memory (Tulving 2002).

5.4.1 Memory of Single Events

Assessing general abilities to remember past events requires data from experiments with many different kinds of stimuli, including stimuli that are not encountered in the wild and are therefore unlikely to engage specialized memory systems. Furthermore, to characterize general memory abilities, negative results about what cannot be remembered are as relevant as results about what can be remembered. The experimental paradigm that best fulfills these criteria is perhaps delayed match-to-sample. In these experiments, the animal first sees a stimulus (the sample), and after a delay it can choose between the sample and another stimulus. It is rewarded if it chooses the sample. Memory is quantified as the percentage of correct choices. Delayed match-to-sample experiments have been conducted many times across taxonomic groups and using different kinds of stimuli, providing us with a representative picture of animal memory for recent stimuli (Macphail 1982, Kendrick et al. 1986, Honig and James 2016). The typical result is that performance decreases sharply as the delay increases by even a few seconds, with only a handful of studies documenting memories lasting up to several minutes (figure 5.2).

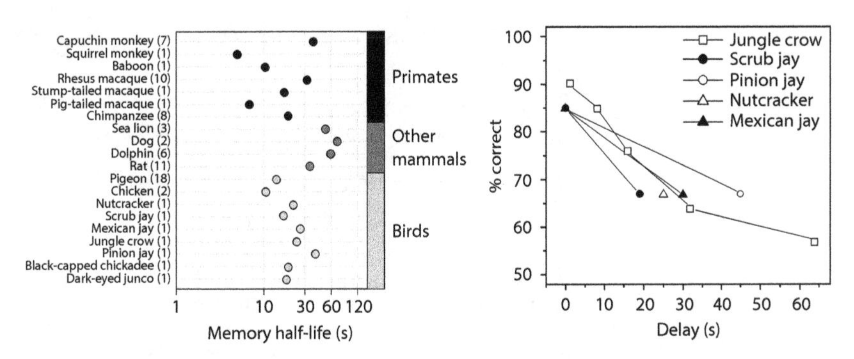

FIGURE 5.2. Memory span in delayed match-to-sample experiments. Left: Data on memory half-life, defined as the delay for which performance drops from its zero-delay value to a value halfway toward chance performance. Estimates are species medians. Numbers next to species names indicate the number of studies. See Lind et al. (2015) for details and data sources. Right: Example of memory decay curves. Data from Goto and Watanabe (2009, jungle crow) and Olson et al. (1995, nutcracker and jays).

Short memory spans have also been observed in primates, in both delayed match-to-sample experiments and delayed response problems (Tomasello and Call 1997, 110). In the latter experiments, the subject sees a food item being hidden in one of two places, and after some delay it is allowed to search for the item. In all primates tested, memory declined substantially within a few minutes. Departing from this pattern is a recent study of dogs (Fugazza et al. 2016), suggesting that some individuals could remember an action performed by a human for up to 24 hours.

Studies of what animals can remember of their own behavior have reached similar conclusions. Rats and macaques, for example, have been trained to alternate between two responses in order to receive rewards, such as pressing one or another lever (Capaldi 1958, Heise et al. 1969, 1976). The animals learned the task as long as they could make responses in rapid succession. Performance degraded sharply, however, when a delay was enforced between one response and the next (figure 5.3). These results indicate that animals have memories of their own behavior, but also that this memory decays quickly—in about a minute in the case of rats and macaques.

In conclusion, available data do not suggest that animals have a general ability to remember events for more than a short time. Rather, the way performance deteriorates in delayed match-to-sample is consistent with a quickly decaying memory trace. Studies of sequences of events provide further support for this conclusion.

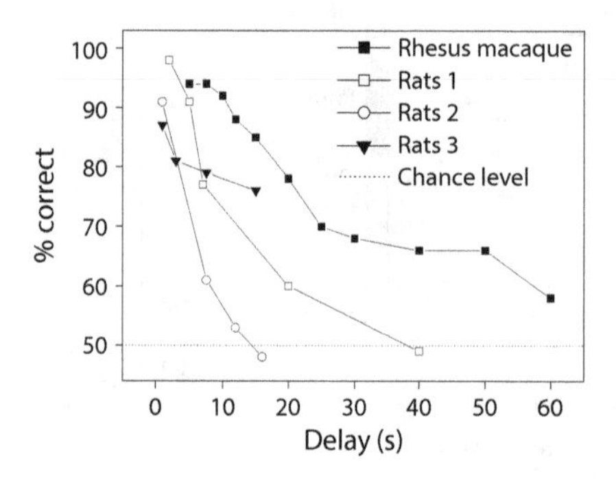

FIGURE 5.3. Performance of rats and rhesus macaques on delayed response alternation experiments. Data from Heise et al. (1976, rats 1), Singer et al. (2009, rats 2), Van Haaren et al. (1985, rats 3), and Brozoski et al. (1979, rhesus macaques).

5.4.2 Memory of Event Sequences

When making decisions, animals can sometimes take into account more than the last perceived stimulus, but this ability seems error-prone and limited to sequences of short duration containing only a few stimuli. For example, Weisman et al. (1980) trained pigeons to respond to a sequence $X \to Y$ and to ignore sequences $X \to X$, $Y \to X$, and $Y \to Y$, where X and Y were colored lights. It took over 500 trials to reach 90% correct responses when discriminating between $X \to Y$ and $X \to X$, while the discrimination between $X \to Y$ and $Y \to Y$ did not reach 90% correct despite about 3000 training trials (figure 5.4, left). In comparison, animals commonly learn discriminations between single stimuli in 25–50 trials. In a later study, Weisman et al. (1985) trained discriminations between sequences of three stimuli, which were found to be increasingly difficult in proportion to the number of identical stimuli at the end of the sequence (figure 5.4, left). The most difficult discrimination, between $Y \to X \to Y$ and $X \to X \to Y$, reached only 55% correct responses in about 3000 training trials. Studies of animals' memory of their own behavior have been less extensive, but have reached similar conclusions. As discussed in the preceding section, animals can learn to choose an action based on their previous action, for example, alternating between left and right lever presses (Mackintosh 1974). It is considerably more difficult, however, to learn tasks in which two or more previous actions must be considered, and at least some errors appear unavoidable (Bloom and Capaldi 1961, Heise et al. 1969, Capaldi 1971, Fountain 2008, Mackintosh 1974).

In a recent review, we concluded that these difficulties are typical of sequence discrimination studies, even when a species' own vocalizations are used to form stimulus sequences, and even in monkeys (Ghirlanda et al.

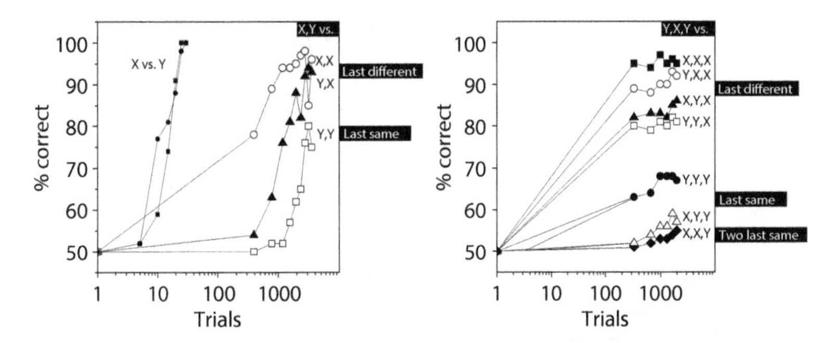

FIGURE 5.4. The difficulty of stimulus sequence discriminations varies greatly. Left: Discrimination of sequences of two stimuli (Weisman et al. 1980, pigeons). For comparison, we also include discriminations between single stimuli in rats: a left-right discrimination in a T-maze (black circles; Clayton 1969) and a black-white discrimination (black squares; Mackintosh 1969). Right: Discrimination of sequences of three stimuli (Weisman et al. 1985, pigeons). For sequence data, each line plots % correct $= 100 \times \frac{a}{a+b}$, where a is responses to the correct sequence (left panel: X, Y; right panel: Y, X, Y) and b is responses to the incorrect sequence.

2017). Read et al. (2022) reach a similar conclusion about working memory in chimpanzees. We also concluded that the same sequence discrimination tasks are much easier for humans, and that nonhuman performance is well described by a model of memory based on the classic idea of stimulus traces illustrated in figure 5.5A (Roberts and Grant 1976, Kendrick et al. 1986). According to our model, animals to do not represent sequence structure faithfully. Rather, a sequence is represented as a set of "traces," each of which describes the memory strength for a stimulus according to its duration and time of offset. For example, in the representation of $X \to Y$, the memory strength of Y is stronger than that of X, because X is no longer present at the end of the sequence and its memory has decayed somewhat. Conversely, in the representation of $Y \to X$, the memory strength of X is stronger (figure 5.5B). We speculated that animals discriminate between sequences based on these differences in memory strengths, and we found this hypothesis in excellent agreement with empirical data. In 65 data sets, we obtained correlations between data and model between 0.58 and 1.00, with a mean of 0.89 (see figure 5.5C for an example). The rates of memory decay we estimated were similar to those estimated from delayed match-to-sample studies (figure 5.2). The memory trace model was accurate even in predicting the performance of songbirds, which might have been predicted to have extraordinary sequence memory (see section 5.5). Our overall conclusion is

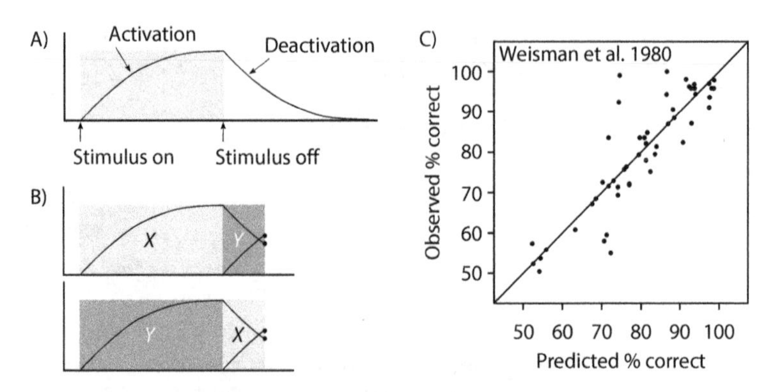

FIGURE 5.5. A: A memory trace. When a stimulus is present, an internal representation of the stimulus builds up; when the stimulus is absent, the representation decays. B: Traces representing sequences X, Y and Y, X can produce similar mixtures of activation, making discrimination difficult. C: A memory trace model accurately predicts the results of Weisman et al. (1980) described in figure 5.4. See Ghirlanda et al. (2017) for details.

that sequence discrimination in animals is typically limited to what a trace memory can accomplish. For example, given the sequence $X \rightarrow Y$, it seems possible to trick animals into thinking that Y came before X by making Y much longer than X, thus resulting in a stronger memory trace for the stimulus that was actually first (MacDonald 1993). Limitations in sequence processing may be important to understand cognitive differences between animals and humans, as sequential information is paramount in human cognition (chapter 9).

5.5 Specialized Memory Systems

The results in section 5.4 about the brevity of animal memory contrast sharply with other observations. Caching birds, for example, can remember for months myriad locations where they have stored food (Bossema 1979, Balda and Kamil 1992, Kamil and Balda 1985), even remembering what they have stored (Clayton and Dickinson 1999a, Sherry 1984) and what food types are rare or abundant at a given time of day (Raby et al. 2007). Caching birds, however, do not have prodigious memory across the board. For example, they fare no better than noncaching birds in delayed match-to-sample experiments (figure 5.2). In fact, the existence of multiple memory systems with different duration is supported by a large body of literature (Sherry and Schacter 1987). One common observation is that animals remember locations and landmarks better than other stimulus features (Gallistel 1990; see also Menzel 1973a, Janson and Byrne 2007). For example, in a pioneering observation, Yerkes and Yerkes (1928) reported that chimpanzees could easily remember the location of food for hours, but had trouble remembering which color indicated food. Some studies of spatial memory are similar to the delayed match-to-sample experiments reviewed in section 5.4, but the animal is required to remember the location of food instead of a visual stimulus. In some of these experiments, rats can remember food locations for at least 24 hours (Ormerod and Beninger 2002), and macaques for at least 25 hours (Beritashvili 1972, Hampton 2001, see also Janson 2016). In another experiment, on learning sets (Zeldin and Olton 1986), rats showed no signs of learning within 1000 trials when nonspatial visual cues were available, but learned within 50 trials with spatial cues (olfactory cues yield even faster learning: Eichenbaum et al. 1986).

There are many other examples of memories for specific kinds of information that are superior in duration, capacity, or other aspects compared to memory for other kinds of events. For instance, animals readily learn to avoid a food they ate many hours prior to falling ill (Garcia et al. 1966, Garcia and Koelling 1966, Garcia et al. 1967), although in other contexts they only associate events separated by at most a few minutes (Mackintosh 1983a). Other examples include male voles remembering for at least 24 hours where they met a sexually receptive female (Ferkin et al. 2008), sexual and filial imprinting in birds and mammals (D'Udine and Alleva 1983, Bolhuis 1991), and song learning in birds (Catchpole and Slater 2003). Imprinting and song learning are examples when experiences, often from an early age, are remembered for very long periods, even for life (Frankenhuis and Walasek 2020).

Studies of episodic memory in animals may provide particularly intriguing examples of memory specialization (Crystal 2010). The concept of episodic memory was first coined to describe the human ability to recount specific events in great detail, such as describing what happened at a particular social occasion (Tulving 1972, 2002). In animals, the term *episodic-like* is used to describe memories that are detailed enough to include at least the what, where, and when of an event, as revealed by suitable experimental tests (Clayton et al. 2007, Crystal 2010, Allen and Fortin 2013). For example, studies of scrub jays indicate that these birds can remember which kind of food they stored at a specific location, as well as the time elapsed since (Clayton and Dickinson 1998, 1999a,b). It is tempting to interpret this ability as deriving from a specialized memory system that has evolved to deal with the demands of a scrub jay's food-caching lifestyle, such as that perishable food items should be retrieved earlier than longer-lasting ones. This interpretation is supported by the fact that scrub jays do not appear to have extraordinary memory in domains other than food caching (see above), and also by the finding that noncaching yet cognitively sophisticated animals, such as chimpanzees, bonobos, orangutans, and gorillas, can fail what-where-when tests of memory (Dekleva et al. 2011, Pladevall et al. 2020).

More generally, multiple memory systems can be seen as embodying knowledge about what kinds of information are relevant to an animal, and over what time span, thereby decreasing learning and memory costs. For example, rabbits can learn to blink upon hearing a sound if the sound precedes an airpuff or electric shock to the eye, but only if the interval between the sound and the noxious stimulus is under 2–3 s (Schneiderman 1966, Mauk and Donegan 1997). Such a short memory span is not necessarily the sign of a rudimentary or "primitive" memory system, but rather of one that is adapted for the short-term purpose of protecting the eye from buzzing insects or other imminent danger. To learn such a protective response, it would be counterproductive to remember what was happening minutes or hours ago. We made similar arguments in chapter 4 when discussing the genetic guidance of learning, of which specialized memory systems are a further example. In conclusion, specialized memory systems tailored to particular needs appear widespread, and play a major role in learning by enabling animals to remember information that is known to be important, based on each species' evolutionary history (Kamil and Roitblat 1985, Gallistel 1990).

5.6 Sources of Information in Associative Learning

The fact that animals take into account a variety of sources of information, including specialized memories, is not in contrast with associative accounts of animal intelligence. An associative learning mechanism can use these sources of information and continue to operate essentially as described in chapter 3. As an example, we consider memory arrangements that can store both long- and short-term information. For example, consider an animal, say, a hummingbird, that forages in a territory with a number of food sources, only some of which are productive at any given time. In this case, it is useful to remember both the location of food sources within the territory and which ones are currently productive. Indeed, hummingbirds remember both the location of flowers and whether they have been recently emptied (Healy and Hurly 1995, Henderson et al. 2001, González-Gómez et al. 2011). This can be accomplished by having two memories that store the same information—one that changes slowly and the other quickly. The slow memory tracks the average value, v, of visiting each location exactly as in the chaining model (we ignore stimulus values, w, for simplicity):

$$\Delta v(S \to B_i) = \alpha \left[u(S') - v(S \to B_i) \right] \tag{5.1}$$

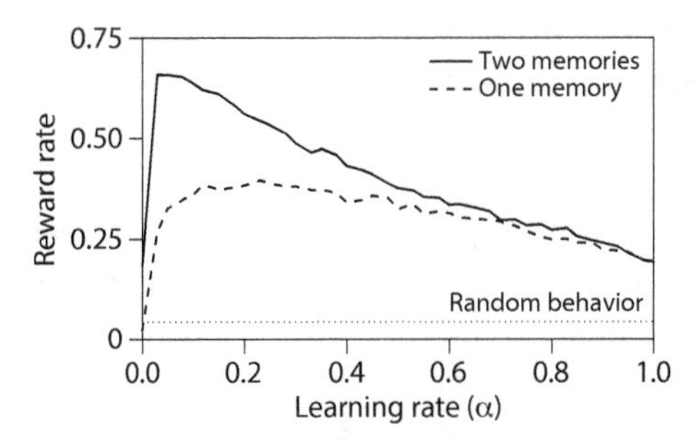

FIGURE 5.6. Simulation of a foraging animal deciding which of 100 locations to visit over repeated foraging trips. Food can be found at 5 locations, which switch between full and empty with a 10% probability between foraging trips. Choosing a location at random has a success rate of 4.5%. The graph compares a one-memory associative learning system (equation 5.1) with a two-memory system that adds a fast memory (equation 5.1 plus equation 5.2), as a function of the learning rate α in equation 5.1. The one-memory system finds food on at most 30% of foraging trips, while the two-memory system can perform twice as well. See section 1.5.5 for simulation details.

where S is a starting location, such as the nest, and B_i the behavior that brings to location i. The reward value $u(S')$ is positive if the location visited is productive (full flower) and zero if it is unproductive (empty flower, or no flower at that location). With a small α, this memory can distinguish between flower and nonflower locations, but does not remember whether a particular flower is full or empty. The latter, however, can be remembered by adding a fast memory, working exactly as in equation 5.1 but with $\alpha = 1$. This memory remembers only the value of the last visit to each location. If we write this value as $x(S \to B_i)$, the corresponding learning equation is

$$\Delta x(S \to B_i) = u(S') - x(S \to B_i) \qquad (5.2)$$

To integrate this memory into decision making, we simply replace $v(S \to B_i)$ with $v(S \to B_i) + x(S \to B_i)$ in the decision-making equation 3.8. With this modification, decision making takes into account both the average value of each location (v) and its last known value (x). Figure 5.6 compares this two-memory system with a single-memory system, as a function of the value of α, in an example with 100 locations, of which 5 are food locations that switch between full and empty with 10% probability on each foraging trip. With a single memory, the best success rate is around 30%, while with two memories it is more than twice as much (see Staddon 2001 for related work on multiple timescales in memory).

The important point of this example is that the two-memory system is still an associative learning system, but it performs in a more sophisticated manner because it has access to more information about the environment. This is true in general: natural selection can tune memory systems to a species' ecological niche, such as selecting one or more appropriate learning rates, tuning which aspects of stimulation are committed to memory, and considering any other source of information discussed above. This generally results in behavior that is more sophisticated than a generic associative learning system with a single memory. However, such sophisticated behavior does not necessarily imply mental mechanisms beyond associative learning. To draw such a conclusion in any specific case, we need to compare the animal's behavior with an associative learning system that has access to the same information the animal is sensitive to.

6

Social Learning

- Social learning occurs when animals learn based on social stimuli, that is, stimuli originating from other individuals.
- In learning models, social stimuli can be treated in the same way as other stimuli.
- Social stimuli can facilitate learning about nonsocial stimuli.
- Associative learning, augmented with genetic predispositions, can account for most of animal social learning, including stimulus and local enhancement, emulation, and at least some cases of imitation.
- With enough training, associative learning may also yield a generalized ability to imitate single behaviors.
- Poor working memory for ordered sequences makes accurate imitation of behavioral sequences difficult for animals.

6.1 Introduction

Social learning can substantially reduce learning costs and enable individuals to learn things they could not learn on their own (Boyd and Richerson 1985, Feldman et al. 1996, Hoppitt and Laland 2013, Aoki et al. 2005). From song learning in birds to tool use in chimpanzees, observations of social learning in animals are plentiful (Heyes 1994, Lonsdorf and Bonnie 2010, Reader and Biro 2010, Zentall 2011, Hoppitt and Laland 2013). This topic is of particular interest to us because animal social learning is often viewed as more sophisticated than associative learning (Tomasello, Kruger, et al. 1993, Whiten 2000, Laland 2004, van Schaik 2010), but also less efficient than human social learning (Lorenz 1973, Galef 1988, Zentall 2006a). In this chapter, we focus on aspects of social learning that we deem relevant to understanding the differences between humans and animals. We begin by summarizing what animals can learn socially in section 6.2. In sections 6.3 and 6.4, we discuss the nature of social information and show that many observations of social learning can result from associative learning. Section 6.5 discusses the limits of associative explanations and the limits of animal social learning. Other aspects of social learning, such as its efficiency in the face of combinatorial dilemmas and its role in supporting traditions and culture, are considered in chapters 12 and 15, respectively.

6.2 What Do Animals Learn Socially?

There are numerous examples of animals learning socially both behavioral skills and preferences for stimuli. Many bird species, for example, learn their song from other individuals (Catchpole and Slater 2003), and brown rats learn to prefer foods that other rats eat (Galef 1996). The benefits of social learning have been demonstrated clearly in many cases. For example, American red squirrels learn to open nuts more easily if they observe an experienced nut opener (Weigl and Hanson 1980), and whooping cranes learn to migrate more efficiently when their flock contains experienced individuals (Mueller et al. 2013). Social learning can also give rise to skills that individuals cannot acquire on their own, such as when black rats learn from their mother to extract seeds from pine cones (Aisner and Terkel 1992, Terkel 1995). Animals also learn from others how to avoid danger: macaque monkeys become fearful of stimuli that make other monkeys afraid, such as snakes (Mineka 1984), and many birds likewise learn to recognize predators from others (Kullberg and Lind 2002, Curio 1988).

Without attempting a comprehensive review, the examples in table 6.1 emphasize that social learning occurs in many behavioral contexts and is taxonomically widespread. Social learning also enables the transmission of behavior from old to young, thus supporting traditions such as song dialects in birds and whales (Marler and Tamura 1964, Rendell and Whitehead 2001; but see Mercado III and Perazio 2021) and tool use in apes (Whiten et al. 1999, van Schaik et al. 2003). The emergence of traditions from social learning has been established in birds, marine mammals, and primates, where specific behaviors have been observed to spread through social interactions (Bonnie et al. 2007, van de Waal et al. 2013, Allen and Fortin 2013, Aplin et al. 2015). In some species, social learning has established multiple traditions and created simple cultures (Heyes and Galef 1996, Laland and Galef 2009, Perry and Manson 2003). For example, we know of more than 60 behaviors that differ across chimpanzee groups (Whiten et al. 1999).

Perhaps the most debated question in the field of social learning is whether animals have the capacity for "true imitation," that is, for learning a behavior through simple observation. True imitation is difficult because of the so-called correspondence problem: the need to translate a stimulus (the *observation* of the behavior) into the corresponding motor act (the *performance*

Table 6.1. Some examples of social learning

Topic	Stimuli	Outcome	References
Food preferences	Conspecific and food smell	Learning new food	Brown rat (Galef 1996)
Food location	Conspecific and novel stimuli	Learning new food	Great and blue tit (Sherry and Galef 1984)
Foraging techniques	Foraging individuals	Operate operant chamber	Pigeon (Zentall et al. 1996)
Migration	Experienced individuals	Learned migratory route	Whooping crane(Mueller et al. 2013)
Antipredation behavior	Warning call or alarm substance	Predator recognition	Blackbird (Curio et al. 1978), minnow (Chivers and Smith 1994), macaque (Mineka and Cook 1988)
Courtship and territorial behavior	Conspecific's song	Song	White-crowned sparrow (Marler and Tamura 1964)
Sexual imprinting	Parents	Partner preference	Finches (Immelmann 1972), zebra finch (ten Cate 1984)

of the behavior; see section 6.3 and Heyes 2001, Brass and Heyes 2005). Because establishing a general correspondence through associative learning appears challenging, it is significant that imitation has been claimed for species as diverse as mammals, birds, reptiles, and octopuses. However, this ability remains a subject of debate given the inconsistency of experimental results. For example, table 6.2 lists 19 experiments, of which 11 reported successful imitation. These studies employed the "two-action" method, in which two groups of animals observed conspecific demonstrators obtaining a reward in two different yet equally effective ways, such as pushing or pulling a lever. Imitation was defined as the observer performing the demonstrated actions significantly more often than the alternative action.

Three aspects of the debate around imitation are especially interesting to our discussion. The first is whether there are animals with a general capacity for imitation, meaning the ability to imitate any action they are capable of executing. Typical two-action studies do not demonstrate such a capacity because they employ simple responses drawn from a species' usual behavioral repertoire. However, dogs have been trained to imitate a wide variety of human actions, demonstrating a capacity not only for generalized imitation

Table 6.2. A selection of studies of imitation using the two-action method

Species	What varies	Description	First trial (%)	Average (%)	Subjects	Source
Octopus	*S*	Red/white ball		**84**	26	Fiorito and Scotto (1992)
Bearded dragon	*B*	Slide door left/right	100	96	8	Kis et al. (2014)
White-eared hummingbird	*S*	Colored flowers	**100**	**100**	20	Lara et al. (2009)
Japanese quail	*B*	Step/peck	**100**	**87.5**	8	Akins and Zentall (1996)
Japanese quail	*B*	Push left/right	**80**	**73**	10	Akins et al. (2002)
Starling	*S* and *B*	Push/pull lid	**100**	**100**	12–18	Campbell et al. (1999) Behavior
Starling	*S* and *B*	Colored plugs		**75**	12–17	Campbell et al. (1999) Stimulus
Budgerigar	*B*	Foot/beak	80	65	10–21	Galef et al. (1986), exp II
Budgerigar	*B*	Foot/beak	50	69	10	Galef et al. (1986), exp I
Budgerigar	*B*	Step/peck		**69**	27	Richards et al. (2009), exp 1
Budgerigar	*B*	Step/peck		**63**	14	Richards et al. (2009), exp 2
Carib grackle	*B*	Open/closed beak	**75**	**69**	20	Lefebvre et al. (1997),* exp II
Pigeon	*B*	Step/peck		**74**	19	Zentall et al. (1996)
Rat	*B*	Push left/right	**88**	**75**	16	Heyes et al. (1994), exp I
Rat	*B*	Push left/right	67	50	18	Heyes et al. (1994), exp II
Common marmoset	*B*	Hand/teeth on lid		**82**	11	Voelkl and Huber (2000)
Common marmoset	*B*	Push/pull door	60	52	5	Bugnyar and Huber (1997)***
Pig-tailed macaque	*B*	Poke/twist		60	11	Custance et al. (2006)
Apes**	*B*	Push/pull door		60	10	Tennie et al. (2006)

Note: Studies are ordered by performance on the first trial after demonstration. Performance is measured as the percentage of observer responses that match the behavior of a demonstrator. Bold indicates performance better than chance. Individuals observed either two different behaviors (*B*) or the same behavior directed to two different stimuli (*S*). The octopus study was the only one to test in extinction. Empty cells reflect missing data.

*2 demonstrator species (Carib grackle and Zenaida dove, *Zenaida aurita*)

**3 orangutans, 1 gorilla, 2 bonobos, and 4 chimpanzees

***Demonstrators only pulled the door

but also for bridging a degree of anatomical difference (Topál et al. 2006, Fugazza and Miklósi 2015). In addition, gray seals have been trained with food rewards to imitate sounds (Stansbury and Janik 2019). The second issue is whether animals imitate actions spontaneously. A study on chimpanzees and human visitors to a zoo suggested that chimpanzees spontaneously imitated the human visitors (Persson et al. 2018), but a later study with better methodology failed to replicate the finding (Motes-Rodrigo et al. 2021). The third issue is whether animals can imitate *sequences* of actions rather than single actions. Most studies in this area have yielded negative results. For example, macaques were unable to reproduce a sequence of two actions performed by another macaque (Ducoing and Thierry 2005, Custance et al. 2006). Negative results have also been obtained for marmosets (Caldwell and Whiten 2004), gorillas (Stoinski et al. 2001), orangutans (Stoinski and Whiten 2003), and chimpanzees (Tomasello et al. 1987, Tomasello, Savage-Rumbaugh, et al. 1993, Myowa Yamakoshi and Matsuzawa 1999, Whiten 1998, Call et al. 2005, Dean et al. 2012). These results are consistent with observations of lengthy and inefficient acquisition of behavioral sequences in nature, such as tool use (section 3.8.4), and with findings that memory for sequences of stimuli is often poor in animals (section 5.4.2). In contrast to this general pattern, Horner and Whiten (2005) found that chimpanzees could imitate a two-action sequence, and the dog studies mentioned above also included short sequences of actions. These issues are relevant for our discussion of the mental mechanisms underlying observations of imitation, in section 6.5.

6.3 The Nature of Social Information

A simple definition of social learning is that it involves some stimuli that originate from other individuals (Heyes 1994, 2012b). We refer to these as "social stimuli" arising from "experienced" animals; but apart from this terminology, we treat social stimuli like any other stimulus. For example, an animal may observe an experienced individual eating a certain fruit and thereby learn that it is worth eating. In this case, the experienced animal and its actions provide social stimuli, while the fruit is a nonsocial stimulus. It is typical of social learning to involve stimulus compounds containing both social and nonsocial elements. We can write such a compound stimulus simply by listing its components,

$$S = S_{\text{social}} S_X S_Y \dots$$

where S is the complete stimulus, S_{social} is a social element of interest, and $S_X S_Y \dots$ represent other stimulus elements, which can themselves be social or nonsocial. In the previous example, S_{social} would come from the observed animal, S_X from the fruit it is eating, and $S_Y \ldots$ from other parts of the environment. Most social learning situations involve a sequence of stimuli with social elements; for example, if the experienced animal performs a sequence such as

$$S \to B \to S'$$

the observer perceives a sequence of stimuli containing social elements, which we write as

$$S_{\text{social}}[S] \to S_{\text{social}}[B] \to S_{\text{social}}[S'] \tag{6.1}$$

where the brackets around S, B, and S' indicate the observer's perception of these stimuli and actions. S and $[S]$ may be very similar, such as the perception of a fruit by the observer and experienced animals, but $[B]$ and $[S']$ are generally very different from B and S'. Namely, $[B]$ is the observer's perception of the experienced animal's behavior, and $[S']$ is the observer's perception of the reward obtained by the experienced animal. If S' is the fruit's sweetness, for example, $[S']$ may be the perception of the experienced animal eating the fruit. The difference between B and $[B]$ implies that behavior established in response to $[B]$ will not spontaneously generalize to B (the correspondence problem, discussed in section 6.2). The difference between S and $[S]$ implies that what is rewarding for the experienced animal is not necessarily rewarding for the observer, which further complicates social learning.

In equation 6.1, we have indicated the general presence of social stimuli with S_{social}, but in more complex situations it may be important to distinguish between different social stimuli, such as the simple presence of another individual or a display indicating danger. In general, social learning occurs if the observer attends to and learns from any of the information available in social stimuli.

6.4 Social Learning through Associative Learning

To make social learning possible, learning mechanisms must extract information from experiences that include social stimuli, store this information in memory, and finally use these memories in decision making. Although there is some skepticism that associative learning can accomplish these tasks (Pearce 2008, Heyes 2012b), several scholars have argued that, in combination with genetic predispositions to react to social stimuli, associative learning can indeed account for many social learning phenomena (Heyes 1994, Huber et al. 2001, Heyes 2012a,b, Miller 2018). To show that this possibility is worth considering, we present here a few concrete examples of social learning in the chaining model (see Lind et al. 2019 for further discussion). Our analysis hinges on the fact that social stimuli can influence what the animal learns about nonsocial stimuli through stimulus values (w if learned or u if innate) and stimulus-response values (v).

To take into account that a compound stimulus often involves social and nonsocial elements, we refine the chaining model slightly as follows. First, we compute the value of a compound simply as the sum of the values of its components. For example:

$$w(S_X S_Y) = w(S_X) + w(S_Y) \tag{6.2}$$

And similarly for $v(S_X S_Y \to B)$:

$$v(S_X S_Y \to B) = v(S_X \to B) + v(S_Y \to B) \tag{6.3}$$

Second, we assume that an experience with a compound stimulus leads to learning about its components. For example, given the experience $S_X S_Y \to B \to S'$, a literal reading of the learning equation for w values (equation 3.11) would yield

$$\Delta w(S_X S_Y) = \alpha_w \left[u(S') + w(S') - w(S_X S_Y) \right] \tag{6.4}$$

Because $w(S_X S_Y)$ is not an independent quantity, however, we assume that the change prescribed by equation 6.4 is applied to $w(S_X)$ and $w(S_Y)$. We assume the same rule for v values. These assumptions are customary in associative learning theory (see Rescorla and Wagner 1972, Ghirlanda 2015, and Enquist et al. 2016 for discussion and refinements). In the following examples, we show how social learning of stimulus values and stimulus-response values can shape behavior.

6.4.1 Social Learning of Stimulus-Response Values

Consider a young bird that approaches conspecifics feeding on a fruit that is unfamiliar to it. This experience can facilitate the bird learning to approach the fruit later on its own, as follows. We assume that the bird has already learned to approach conspecifics and to find the presence of conspecifics rewarding, or that it has genetic predispositions to this effect. In the chaining model, if S_{social} is the presence of conspecifics and B is the approach behavior, we have $u(S_{\text{social}}) > 0$ and $v(S_{\text{social}} \to B) > 0$. The bird's experience can be written as

$$S_{\text{social}} S_X \to B \to S' \tag{6.5}$$

where S' is the stimulus perceived after approaching the conspecifics. Before we consider S' in detail, note that, following the learning rules

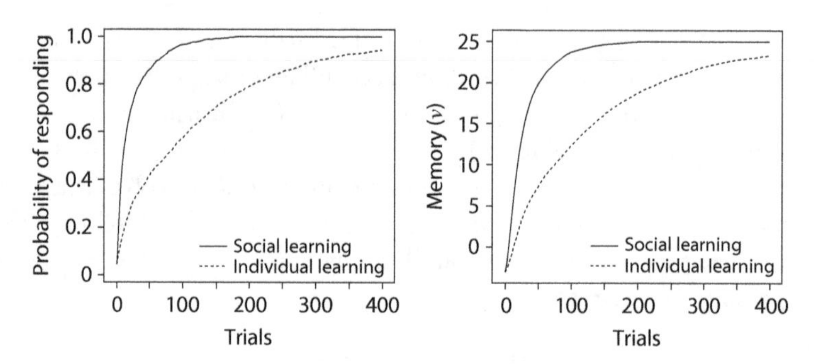

FIGURE 6.1. Learning a response to a novel stimulus through social or individual learning. The learner is exposed repeatedly to the stimuli $S_{soc}S_X$ (with probability 0.2) and S_X (with probability 0.8) in the social learning case, and only to S_X in the individual learning case. Left: Probability of responding toward S_X under social and individual learning. Right: Underlying changes to associative memories. Primary values of stimuli are $u(S') = 25$ and $u(S_{soc}) = 3$. See section 1.5.5 for simulation details. Curves are the average of 1000 subjects.

introduced above,

$$\Delta v(S_X \to B) = \alpha_v \left[u(S') + w(S') - v(S_{social} \to B) - v(S_X \to B) \right] \quad (6.6)$$

According to this equation, $v(S_X \to B)$ increases if $u(S') + w(S')$ is larger than $v(S_{social} \to B) + v(S_X \to B)$. Note now that S' represents what the young bird perceives after having approached the conspecifics. Hence S' still contains S_{social} and is therefore perceived as valuable. Because we have assumed that $v(S_X \to B)$ is low (the bird is not approaching the fruit), it is enough that $v(S_{social} \to B)$ not be too large for equation 6.6 to result in an increase in $v(S_X \to B)$. Thus approaching a compound stimulus consisting of conspecifics eating the fruit has made the young more likely to approach the fruit on its own. Figure 6.1 simulates the effect of repeated experiences, illustrating that social learning leads to a more rapid acquisition of approach behavior than individual learning.

In this example, the approach behavior in response to S_{social} is also the appropriate behavior in response to the unknown stimulus, S_X. However, additional behaviors are often required to respond fully to S_X (in this case, eating the fruit). Learning these additional behaviors may occur through simple trial and error, but predispositions involving S_{social} can help. For example, seeing a conspecific feeding not only would reward approach but could also limit exploration to behaviors relevant to feeding. This is an example of a context-dependent repertoire, as discussed in chapter 4. In conclusion, by beginning with the assumption that social stimuli can reward behavior, we find that associative learning can result in social transmission of a behavior or even a short behavioral sequence.

6.4.2 Social Learning of Stimulus Values

If we also assume that social stimuli can influence the value of other stimuli, associative learning becomes even more capable of supporting social learning. In the previous example, approaching a feeding conspecific would not only increase the tendency to approach the fruit but would also endow the fruit with stimulus value. That is, experiencing the sequence in equation 6.5 would make both $w(S_X)$ and $v(S_X \to B)$ positive, in turn rewarding approach of the fruit independent of the presence of conspecifics. Here we explore this effect and link it with several social learning phenomena. Suppose that the goal of the observer is to learn a sequence of two behaviors,

$$S_X \to B_1 \to S_Y \to B_2 \to S' \tag{6.7}$$

where S' is a reward. Following our previous example, this sequence might be

$$\textit{Fruit far} \to \text{Approach} \to \textit{Fruit near} \to \text{Eat} \to \textit{Sweet}$$

When the sequence is performed by another animal, it is perceived by the observer as

$$S_{\text{social}}S_X \to S_{\text{social}}[B_1] \to S_{\text{social}}S_Y \to S_{\text{social}}[B_2] \to S_{\text{social}}[S'] \tag{6.8}$$

where we assume, for simplicity, that S_X and S_Y are perceived similarly by both individuals. The stimulus S_{social} is a placeholder for any social stimuli, which can change throughout the sequence. For example, the last S_{social} could indicate the observed animal's reaction to eating the fruit, which the observer may interpret as rewarding.

The sequence given by equation 6.7 is not too complicated to learn individually, yet social learning experiences of the form shown in equation 6.8 can dramatically speed up learning, as seen in figure 6.2. The reason is that the social learning situation exposes the learner to the whole sequence and thus provides a more favorable entry pattern (section 3.5). We can illustrate this effect by comparing an individual learner and a social learner, both learning the sequence in equation 6.8. According to the chaining model, the individual learner would initially only experience S_X, and would experience S_Y and the reward S' only after stumbling by chance upon the correct behaviors, B_1 and B_2. The initial performance of B_1, in particular, would be unrewarded. The social learner, on the other hand, would immediately experience S_X, S_Y, and $[S']$ in the correct order, even without performing B_1 and B_2. If $S_{\text{social}}[S']$ is

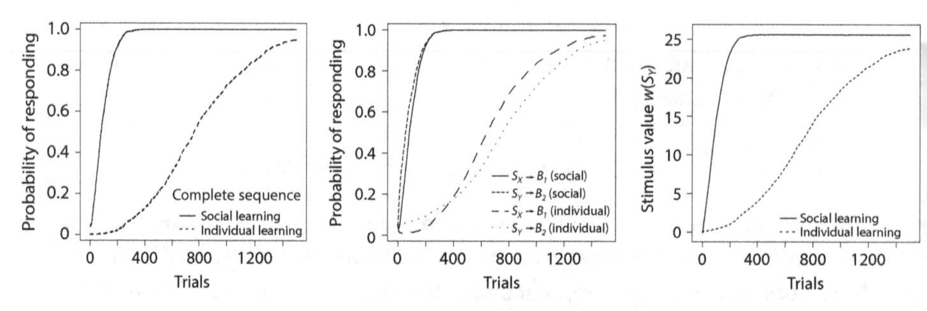

FIGURE 6.2. Social learning of a sequence of two behaviors through chaining (equation 6.7). The social learner is exposed repeatedly to $S_{\text{social}}S_X \to S_{\text{social}}S_Y$ (with probability 0.2) and S_X (with probability 0.8); the individual learner is exposed only to S_X. Left: Learning curves for the acquisition of the complete sequence. Middle: Probability of correct responses toward S_X and S_Y. Right: Stimulus value (w) of S_Y, which acts as a conditioned reinforcer of the $S_X \to B_1$ response. Simulation parameters as in figure 6.1. See section 1.5.5 for simulation details. Curves are the average of 1000 subjects.

rewarding, S_Y accrues stimulus value, which in turn rewards responding to S_X with B_1. Thus the social learner is rewarded earlier than the individual learner for performing B_1, which speeds up learning of the whole sequence. The same argument applies to sequences of more than two behaviors.

Associative learning can also result in avoidance behavior through a similar mechanism of value transfer from social to nonsocial stimuli (Mineka 1984). For example, a naive monkey that observes a conspecific reacting in fear to a snake may assign a large negative value to the snake stimulus, which encourages avoidance of snakes. We expect genetic predispositions to play an even larger role in social learning of avoidance than of appetitive behaviors, in part because learning to avoid danger must be rapid and reliable. Thus monkeys learn to fear snakes much more quickly than they learn to fear innocuous objects (Cook and Mineka 1990).

Many social learning phenomena can be explained by taking into account that social situations provide animals with favorable entry patterns, whereby they experience stimuli they would not have otherwise experienced. These stimuli, moreover, can acquire value and therefore guide the animal's actions toward productive behavior.

Stimulus enhancement and *local enhancement* are usually described as direction of the observer's attention to a particular stimulus or location, allowing learning of a straightforward behavior, such as approach or avoidance (Heyes and Galef 1996). A classic example comes from small birds learning to open milk bottles to steal the cream floating on top (Fisher and Hinde 1949).

This behavior spread throughout England as inexperienced birds learned to approach milk bottles by observing experienced birds foraging at the bottles. Each bird, however, learned on its own to pierce the foil seal and retrieve the cream (Zentall 2004). Rather than describing the phenomenon of enhancement as "drawing attention" to stimuli or locations, we can define it as the transfer of value from social to nonsocial stimuli. This could serve to reward approach, avoidance, or other behaviors.

Emulation refers to an attempt to reproduce the outcome of an action without imitating the action itself (Tomasello, Savage-Rumbaugh, et al. 1993, Tomasello 1996, Call et al. 2005, Tennie et al. 2010). For example, from observing others, a chimpanzee may understand that nuts must be cracked before they can be eaten, but might be unable to copy the nut-cracking sequence. Our interpretation of this learning situation is that intermediate stimuli, such as open and cracked nuts, can acquire value when accompanied by social stimuli, which then help the individual learn nut cracking. Both observations of nut cracking in nature (Inoue-Nakamura and Matsuzawa 1997) and our analysis of it in section 3.8.4 and Enquist et al. (2016) show that mastering the technique is very time consuming. An unaided individual is not likely to persist, but social stimuli may encourage perseverance.

Structured situations of emulation provided by a parent or other conspecific are sometimes called *opportunity providing* (Ewer 1969, Caro and Hauser 1992, Hoppitt et al. 2008). For example, many cat species provide live prey for their young to practice hunting and predatory behavior (Ewer 1968). Opportunity providing is sophisticated enough in meerkats that we might even call it teaching. The parent initially brings dead scorpions to the young, then live but disabled scorpions, and finally intact scorpions, so that the young improve their predatory skills gradually (Thornton and McAuliffe 2006). In this case, the role of the social situation in providing a favorable entry pattern to a difficult task is particularly clear.

6.5 Can Associative Learning Account for Animal Imitation?

In the previous section, we have seen that associative learning can give rise to various social learning phenomena. In this section, we discuss whether it can account for imitation, the most sophisticated form of social learning. Both chaining and other associative learning theories suggest only limited observational learning of this kind.

We discuss chaining first. Consider again the social learning sequence in equation 6.1. The observed animal performs

$$S \to B \to S'$$

while the observer perceives

$$S_{\text{social}}S \rightarrow S_{\text{social}}[B] \rightarrow S_{\text{social}}[S'] \tag{6.9}$$

where we have assumed, for simplicity, that the nonsocial stimulus, S (e.g., a fruit or other object), is perceived similarly by the observer and the observed. Following equation 6.9, successful imitation requires the observer to increase the stimulus-response value, $v(S \rightarrow B)$, which is the variable that underlies the performance of B in the presence of S. Effecting this change, however, requires solving both a correspondence problem (section 6.2) and a reinforcement problem. Namely, to increase $v(S \rightarrow B)$, the observer must first understand that the observation $[B]$ corresponds to its own behavior B (correspondence problem), and then it must to update $v(S \rightarrow B)$ even though the experience in equation 6.8 has resulted in no direct reward (reinforcement problem). We have seen that attributing value to social stimuli may solve the reinforcement problem, but the correspondence problem remains. Most studies of imitation have employed behavioral responses that are likely genetically predisposed (Heyes 2001), such as pecking and scratching in foraging birds (table 6.2). Learning these predisposed responses does not require solving the correspondence problem because the responses are spontaneously tried, and we have argued above that chaining provides various ways to "imitate" when restricted to such responses (section 6.4.1). However, more general imitation abilities cannot be explained thus (e.g., in dogs: Topál et al. 2006, Fugazza and Miklósi 2015). Other associative learning theories, while differing in some respects from chaining, paint much the same picture. For example, observational learning has been conceptualized as Pavlovian conditioning of stimulus-stimulus associations (Heyes 2012b). However, Pavlovian conditioning is not a general solution to the correspondence problem because it affects behavior only through genetically predisposed responses with limited flexibility (Bouton 2016). Heyes (2012b) has suggested that observational learning of stimulus-response associations is also possible, but this ability still implies solving the correspondence problem.

Can the correspondence problem be solved by learning? Interest in this possibility has been rekindled (Heyes 2001, 2010) despite earlier suggestions that imitation must rely on an innate stimulus-to-response translation (e.g., through mirror neurons: Rizzolatti and Craighero 2004). Indeed, it appears that even humans have to learn to imitate (Heyes 2016, Oostenbroek et al. 2016). We may speculate that associative learning can result in generalized imitation if extensive feedback is provided for many behaviors, such that eventually the animal learns to map a multitude of stimuli to their corresponding behaviors. This is similar to how imitation has been trained in dogs (Fugazza and Miklósi 2015). Imitation of novel behaviors could then arise by stimulus generalization—that is, because the novel behaviors share perceptual elements with familiar behaviors that have a known stimulus-to-response translation. If this hypothesis is correct, we might even speculate that general-purpose imitation does not routinely arise in animals, not because they lack the necessary learning skills but simply because they are not sufficiently "trained" in their natural environment. For example, Bering (2004) has suggested that imitation in chimpanzees arises as a result of human training. Species with larger behavioral repertoires would require more training to imitate reliably, which may be why primates do not stand out as exceptional imitators (table 6.2).

Imitating more than a single behavior—that is, a behavioral sequence—requires remembering long stimulus sequences faithfully. In chapter 5, we concluded that animals lack such a general sequence memory, making imitation of behavior sequences in animals very difficult in the absence of strong genetic support.

7

Can Animals Think?

- Chaining differs from thinking because it does not recombine information from different events, and does not store information about causal relationships between events.
- However, chaining and thinking have been difficult to distinguish in practice.
- We review a diversity of data about the possibility of causal learning and thinking in animals, including experiments on outcome revaluation, latent learning, route planning, and insights.
- Most results fit associative learning at least as well as thinking, and sometimes better.
- Future research should focus on situations in which chaining and thinking make more distinctive predictions, including
    - Learning of behavioral sequences rather than single behaviors
    - Distinguishing between conditioned and primary reinforcement
    - Time-consuming decision making
- Unambiguous predictions will require theories for causal learning and thinking to be formalized mathematically.

7.1 Introduction

Up to now we have described animal learning as chaining (chapter 3) guided by genetic predispositions (chapter 4), supported by both domain-general and domain-specific memories (chapter 5). In this chapter, we consider the evidence for more sophisticated mental mechanisms; in particular, mechanisms that would qualify as "thinking," "insight," or "mental problem solving." In section 2.5, we defined thinking as the recombination of causal information gathered from different experiences, a definition that also echoes common uses of insight, mental problem solving, and related terms. We start this chapter by considering some predictions drawn from this definition, as well as challenges in testing such predictions (section 7.2). The predictions concern the extent to which animals can change their behavior upon receiving new information (sections 7.3 and 7.5), whether they gather causal information of no immediate value (sections 7.4 and 7.6), and whether animals engage in both fast and slow decision making, with the latter suggesting internal processes more complex than associative learning (section 7.7). In addressing these predictions, we consider experiments, such as finding shortcuts in mazes or demonstrating "insight" into problems, that have been put forward as evidence for recombination of information beyond chaining. However, we conclude that the evidence is compatible with, or even supportive of, chaining (see also Shettleworth 2010a, Heyes 2012a), and that more controlled experiments are needed to make progress on this issue (sections 7.8 and 7.9).

7.2 The Recombination of Causal Information

There are three key properties of thinking mechanisms that derive from our definition of thinking in section 2.5. These properties can guide our analysis of empirical data, because chaining exhibits none of them.

First, in a thinking mechanism, experiences with a stimulus can have immediate impact on behavior in response to perceptually unrelated stimuli. Suppose you enjoy pizza and often enter a local shop to buy some. One evening, you order delivery from the shop and fall ill with food poisoning, causing pizza to become strongly aversive. If you had learned your pizza-procuring behavior purely by chaining, you might still enter the pizza shop, because it had previously been rewarding; only after seeing the pizza would you realize that entering the shop has negative consequences. If you can think, on the other hand, you can draw the conclusion that it's best not to enter to buy a pizza that you would not enjoy. Do animals behave as if drawing such conclusions? We consider two kinds of tests: revaluation experiments, which are similar to our pizza example, and spatial rerouting experiments, in which animals must plan a novel route when a known route becomes inaccessible.

Second, a thinking mechanism would gather causal information even when its value is uncertain, in anticipation of possible future relevance. For example, you may notice a new pizza restaurant when you are not hungry, and return to that location later when you feel like having pizza. Even this mundane behavior requires storing information (the location of the new restaurant) without any immediate benefit for doing so. Chaining, however, only learns when the animal perceives a difference between the experienced and estimated stimulus values or stimulus-response values. Do animals learn even without perceiving such differences? This question has been considered extensively since the early days of experimental psychology, and we discuss several examples.

Third, thinking can take time because it may require many mental steps. Consider a city with a large subway system, such as New York, London, or Paris. Even if you know the subway map by heart, it will take some time to figure out the best route between two unfamiliar destinations. Chaining would not be able to arrive at a solution without actually trying out the alternatives, but humans can mentally compare different alternatives given some time. Is there evidence of time-consuming decision making in animals, suggestive of similar mental processing? This topic is less researched than the first two. We consider a few studies to stress its potential importance.

Demonstrating recombination of causal information in animals poses several practical challenges. For example, that an animal can learn a complex task is not per se evidence of thinking, as chaining can also learn complex tasks (chapter 3). Chaining and thinking, however, often differ over the course of learning. Namely, we expect thinking mechanisms to learn faster (provided they have sufficient information) and to avoid mistakes that would arise from previously established behavior, such as entering the pizza shop when you would not eat pizza. For these reasons, few conclusions can be drawn unless the animals' history of experiences is adequately known (Lind et al. 2009, Lind 2018, Barker and Povinelli 2019). Thus the most informative studies are likely to be laboratory experiments that start with naive animals and that document behavior thoroughly during both training and testing.

7.3 Revaluation of Past Experiences

Most data on the revaluation of past experiences come from experiments that are similar to our pizza example. The most common one is, in fact, a *devaluation* experiment in which a positive outcome is turned into a negative one, although some experiments have increased reward value. The prototypical experiment has a training phase, a revaluation phase, and a test phase, in that order. In the training phase, animals learn to perform a sequence, such as

$$S_1 \to B_1 \to S_2 \to B_2 \to S_{\text{outcome}} \tag{7.1}$$

some concrete examples of which are shown in table 7.1. Typically, the first part of the sequence ($S_1 \to B_1 \to S_2$) is trained during the experiment, while the second part ($S_2 \to B_2 \to S_{\text{outcome}}$) is already established when the experiment starts. In the revaluation phase, the animals experience that the consequences of B_2 in response to S_2 have changed:

$$S_2 \to B_2 \to S_{\text{revalued}} \tag{7.2}$$

Table 7.1. Behavioral sequences trained in the experiments in figure 7.1

S_1	→	B_1	→	S_2	→	B_2	→	$S_{outcome}$	$S_{revalued}$
Runway	→	Run	→	*Flavored water*	→	Drink	→	*Hydration*	*Hydration* + *toxin*
Lever	→	Press	→	*Food*	→	Eat	→	*Nutrients*	*Nutrients* + *toxin*
Lever	→	Press	→	*Flavored water*	→	Drink	→	*Hydration*	*Hydration* + *sucrose*

where $S_{revalued}$ is a stimulus with a different value than $S_{outcome}$. Finally, in the test phase, the animals are presented again with S_1, and their propensity to perform B_1 is measured. The test is performed in extinction, that is, with B_1 having no consequence. In the most common realization of this design, rats are trained to press a lever for food, then they are given the food with a toxin to decrease its value, and finally they are presented with the lever again to see whether they press it or not.

If the reward has been devalued, an animal capable of recombining training and revaluation experiences should be reluctant to perform B_1, understanding that now the sequence will have a negative outcome. If the reward value has been increased, however, recombination of information would lead the animal to perform B_1 more often. In both cases, an animal that learns by chaining is predicted to perform B_1 at the same rate as before, at least initially, because the revaluation experiences did not involve S_1 and thus should not have modified the stimulus-response value $v(S_1 \rightarrow B_1)$ that underlies the performance of B_1. Performance of B_1 is usually compared to that of control subjects, such as animals that undergo the training and test phases but not the devaluation phase.

7.3.1 Do Revaluation Experiments Show Recombination of Information?

Figure 7.1 illustrates the variety of results obtained in revaluation experiments. The left panel is from Chen and Amsel (1980), in which thirsty rats were trained to traverse a runway in order to reach flavored water. The water was then devalued with a toxin, and later the rats were placed back in the runway without a water reward. Experimental animals ran at slower speeds compared to controls that had received no devaluation treatment. The difference in running speed was apparent from the very first trial and thus cannot be attributed to learning during the test. Moreover, devaluation took place in the rats' home cages so that the experimental animals could not have associated illness with the runway. These results are consistent with the

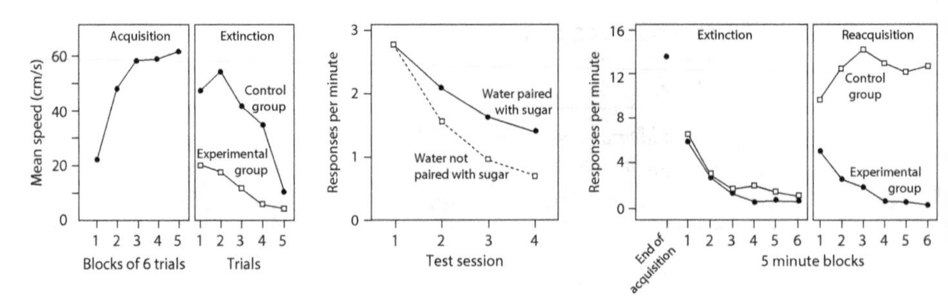

FIGURE 7.1. Possible results of revaluation tests in extinction. Left: Revaluation evident from the first test trial (Chen and Amsel 1980). Middle: Revaluation emerging during testing (Rescorla 1990, experiment 1). Right: No revaluation in extinction (Adams 1980). We show data only from the experimental group, which received the revaluation treatment, and a control group, which did not receive the treatment. Some studies included additional control groups.

rats recombining information from training and devaluation experiences to conclude that traversing the runway would lead to a negative outcome.

The middle panel in figure 7.1 shows results by Rescorla (1990, experiment 1). In this case, rats were initially trained to press a lever for plain water, after which the value of the reward was increased by the addition of sugar. If the rats could recombine training and revaluation experiences, they should now press more often than control rats. This increase in responding was not immediately apparent, but it developed during testing: when lever pressing yielded nothing, experimental rats proved more persistent in pressing than control rats. We call this *delayed* revaluation, to distinguish it from the *immediate* revaluation observed by Chen and Amsel (1980). Note that the test was done "in extinction," that is, with lever pressing yielding no food. This is common, and important, to make sure that information about the changed value of the food can only come from the devaluation experiences rather than from test experiences.

The right panel in figure 7.1 shows results from Adams (1980, experiment 1), in which rats were trained to press a lever for a food reward, and then the food was devalued by pairing it with a toxin. Devaluation occurred in the same experimental setting but in the absence of the lever, so that the rats could not perform the trained response. When the lever was made available again (without yielding any food), experimental rats pressed less than controls throughout the extinction test, thus showing no revaluation effect. A stark effect, however, was present in the subsequent reacquisition tests, in which control animals resumed pressing when the food was made available again, but experimental animals did not.

Table 7.2 summarizes a broader sample of revaluation studies, confirming that results vary greatly. In devaluation studies, immediate or delayed revaluation is common, whereas studies in which reward value is increased have yielded mixed results (table 7.1). It is unclear whether there is an inherent difference between reward increase and decrease, or perhaps between poisoning and other revaluation techniques, or whether decreasing value is simply easier. Revaluation has also been attempted by changing the animal's motivational state, such as hunger level. In these cases, the most common result is no revaluation or delayed revaluation. For an example, see our discussion of Spence and Lippitt (1946) in section 7.4.3. We also note that very similar experiments can yield contrasting results. For example, Dickinson et al. (2002) found revaluation with food but not ethanol rewards, and Colwill and Rescorla (1985) found stronger effects with solid food than with a sucrose solution.

In summary, revaluation experiments have widely varying outcomes. When revaluation is effective, behavioral change is often delayed rather than immediate. This variability suggests that revaluation may not be the outcome of mechanisms that can freely recombine information. Rather, recombination may be specific to some stimuli, contexts, and procedures (cf. section 5.6). Particularly strong evidence against recombination comes from experiments in which animals are trained to perform a sequence of two actions in order to obtain a reward, such as pressing a lever and then pulling a chain. If the second action is extinguished on its own, no longer resulting in the reward, a mechanism capable of recombination would conclude that the first action has become less valuable. However, extinguishing the second action typically has little or no effect on the first (Skinner 1934, Thrailkill and Bouton 2016). Indeed, synthesizing evidence from relevant studies, Balleine and Dickinson (2005) suggest that revaluation initially affects only the final action(s) of a behavioral sequence, whereas earlier actions are extinguished only after multiple executions of the sequence. In other words, information about the altered reward value would travel backward along a behavior sequence rather than affecting the whole sequence at once, in agreement with chaining.

7.3.2 Can Chaining Account for Revaluation Findings?

In this section, we consider chaining as a candidate explanation for the results reviewed above, analyzing the typical devaluation design in table 7.3. This experiment comprises (1) an acquisition phase, in which lever pressing for food is trained, (2) a devaluation phase, in which a toxin is administered following food consumption, (3) an extinction phase, in which lever pressing yields no food, and (4) a reacquisition phase, in which lever pressing again yields food. For brevity and generality, we use the labels S_1, S_2, B_1, and B_2

Table 7.2. A sample of revaluation studies with rats

	Reward ($S_{outcome}$)	Treatment ($S_{revalued}$)	Revaluation
Reward devaluation			
Tolman (1933)	Food	Shock	None / immediate*
Skinner (1934)	Food	Extinction	Delayed
Miller (1935)	Food/water	Shock	Immediate†
Tolman and Gleitman (1949)	Food	Shock	Immediate*
Rozeboom (1957)	Water	Extinction	None
Adams (1980)	Food	Food + toxin	Delayed
Chen and Amsel (1980)	Water	Food + toxin	Immediate (2), none (1)
Adams and Dickinson (1981)	Food	Food + toxin	Within ~ 20 responses
Colwill and Rescorla (1985)	Food	Food + toxin	Within 2–8 responses
Colwill and Rescorla (1985)	Sucrose	Sucrose + toxin	Within ~ 10 responses
Rescorla (1990)	Water	Water + toxin	Delayed
Rescorla (1992, 1994)	Food	Food + toxin	Within 5–10 responses
Dickinson et al. (2002)	Food	Food + toxin	Within ~ 5 responses
Dickinson et al. (2002)	Ethanol	Ethanol + toxin	None
Thrailkill and Bouton (2016)	Food	Extinction	>10
Reward inflation			
Seward (1949)	None	Food	Immediate (2)
Young et al. (1967)**	None	Food	Immediate (1), none (1)
Rescorla (1990)	Water	Water + sucrose	Delayed
Motivational change			
Spence and Lippitt (1946)	Food/water	Reverse motivation§	None
Kendler (1947)	Food/water	Reverse motivation§	None
Balleine (1992)	Food	Increase hunger	Delayed
Colwill and Rescorla (1985)	Food	Decrease hunger	Within 4–8 responses
Dickinson and Burke (1996)	Food	Decrease hunger	None‡, delayed
Balleine (1992)	Food/sugar water	Increase hunger	None, delayed
Shipley and Colwill (1996)	Food/sugar water	Increase hunger	None, delayed
Balleine and Dickinson (1998)	Water	Decrease thirst	Delayed

Note: Revaluation is labeled "immediate" if apparent from the first test trial, "delayed" if it becomes apparent during testing, and "none" if no effect is apparent. In the absence of sufficient details to conclude whether revaluation was immediate or delayed, we estimated the number of responses within which revaluation was apparent. Numbers in parentheses in the Revaluation column indicate how many experiments gave the indicated result.
*Compared to random performance rather than to a control group
**Also tested squirrel monkeys, finding no effect
†Compared to baseline performance rather than to a control group
‡Result stated verbally with no accompanying data
§Treatment involved shifting animals from thirsty to hungry

Table 7.3. A devaluation experiment

Phase	Experiences								
	S_1	→	B_1	→	S_2	→	B_2	→	$S_{outcome}$
1. Acquisition	*Lever*	→	Press	→	*Food*	→	Eat	→	*Nutrients*
2. Devaluation					*Food*	→	Eat	→	*Nutrients* + *Illness*
3. Extinction	*Lever*	→	Press	→					*Nothing*
4. Reacquisition	*Lever*	→	Press	→	*Food*	→	Eat	→	*Nutrients*

for relevant stimuli and behaviors, as indicated in table 7.3. In our analysis, we are mainly interested in tracking the stimulus-response value, $v(S_1 \to B_1)$, which influences the performance of the focal behavior, and the stimulus value, $w(S_2)$, which influences $v(S_1 \to B_1)$ in the chaining model because S_2 is experienced after $S_1 \to B_1$.

The basic predictions of the chaining model are as follows. At the end of acquisition, both $v(S_1 \to B_1)$ and $w(S_2)$ have a high positive value, corresponding to the animal pressing the lever and valuing the food. In the devaluation phase, $w(S_2)$ becomes negative because S_2 is now followed by a negative outcome. In this phase, however, B_1 is never executed and thus $v(S_1 \to B_1)$ does not change. Consequently, in the extinction phase, chaining predicts the same behavior regardless of whether the devaluation phase is included, as shown in the left panel of figure 7.2. During reacquisition, however, the sequence $S_1 \to B_1 \to S_2$ is experienced, and thus the negative $w(S_2)$ acts to depress $v(S_1 \to B_1)$. This leads to quick extinction of the response $S_1 \to B_1$ and hinders the reacquisition of the response. If devaluation is omitted, however, the response is promptly reacquired, as shown in figure 7.2.

The predictions just spelled out match the results from Adams (1980), as shown in the right panel of figure 7.1. However, as we have seen above, other results are possible (table 7.2). Chaining can produce revaluation during extinction, if we assume that some stimuli that acquire a negative w value during revaluation are also present during extinction. For example, in some studies, the animals consumed the food in the same apparatus used during acquisition and extinction. If we label as X stimuli from the apparatus that are common to all phases of the experiment, such as the food receptacle, we can conceptualize the devaluation experiences as

$$\textit{Food} + X \to \text{Eat} \to \textit{Nutrients} + \textit{Illness}$$

We can then see that both $w(\textit{Food})$ and $w(X)$ would become negative. Furthermore, the apparatus would also be experienced during extinction, as

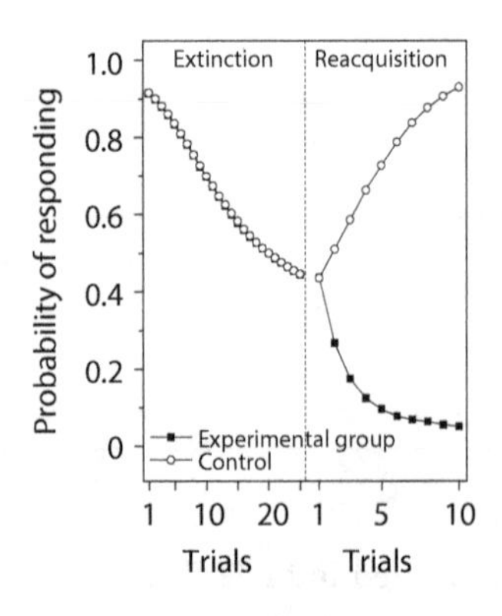

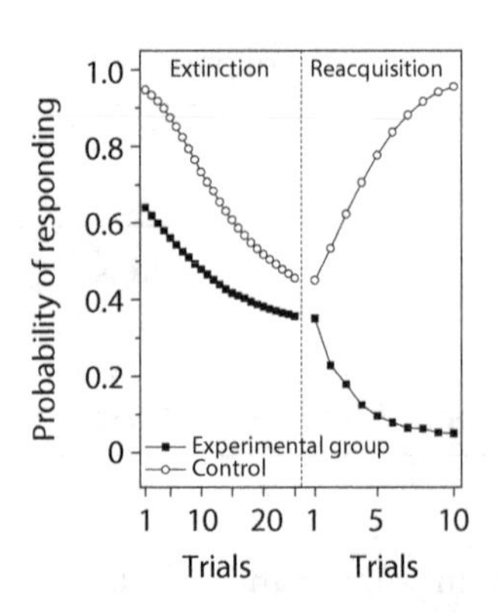

FIGURE 7.2. Two simulations of the chaining model of revaluation experiments in which an animal is first trained to press a lever to obtain a food pellet, then given a single experience with a poisoned pellet, and then given an extinction test (the lever is presented, but pressing it has no consequence) and a reacquisition test (lever pressing again produces unadulterated food pellets). The difference between the simulations is that the right panel included an additional trial before extinction testing, in which the model experienced the whole chain *Lever* → Press lever → *Pellet*, allowing the negative value of the pellet to backtrack to the lever. Learning parameters: $\alpha_v = \alpha_w = 0.1, \beta = 1, u(\mathit{Nutrients}) = 10$, $u(\mathit{Illness}) = -300$. The cost of actions Press and Eat was set to 1, while not responding was cost-free. Curves are the average over 1000 subjects. Training prior to the test consisted of 25 trials, resulting in a frequency of lever pressing above 0.9. See section 1.5.5 for simulation details.

the rat looks for food after a lever press. Thus the extinction phase includes experiences of the form

$$\mathit{Lever} \to \text{Press} \to X$$

which provides an opportunity for the negative $w(X)$ to decrease lever pressing. This leads to delayed revaluation.

In summary, chaining may account for the most common revaluation findings: no revaluation and delayed revaluation. Immediate revaluation may be harder to explain in terms of associative learning, but not impossible. For example, when devaluation takes place in the same apparatus in which the instrumental response is performed, the animal may learn to avoid parts of the apparatus, which may later interfere with the response. This and other associative explanations of immediate revaluation are discussed by Miller (1935) and Adams and Dickinson (1981). A distinct possibility is that many instances of "immediate" revaluation result from averaging over many responses. For example, Adams and Dickinson (1981, experiment 1) reported a revaluation effect in the first data point of their revaluation test. This data point, however, included at least 20 responses—meaning that the published data are, in fact, compatible with both delayed and immediate revaluation. The only behavioral parameter that is completely immune to this confound is probably the time until the first response is performed. Even if all these explanations are rejected, the rarity of immediate revaluation indicates that mechanisms beyond chaining would only be engaged under very specific circumstances.

7.3.3 Stimulus Value and Revaluation

Our analysis suggests that it may be possible to account for revaluation results through learned stimulus values, rather than through mental mechanisms that recombine information from different experiences. The potential role of stimulus values is demonstrated elegantly by Parkinson et al. (2005). The crucial aspect of this experiment is that stimulus value was manipulated explicitly rather than incidentally, as, we have hypothesized, occurs in revaluation experiments. Thus, in Parkinson et al.'s experiments, rats first learned that a light was valuable as it reliably signaled the delivery of sweetened water. The latter was then devalued by inducing sickness, resulting in a substantial decrease in consumption. At this point, the rats were given the opportunity to press a lever in order to turn on the light, which they did as much as control animals for which the sweetened water had not been devalued. Thus lever pressing was sustained solely by the learned value of the light. The devaluation procedure did not affect this value, even if it had been learned based on an outcome that had become undesirable. This result agrees with the chaining model, in which the value $w(Light)$ is not changed by experiences with the outcome. On the other hand, the result disagrees with a model in which the animals can reason based on facts, such as "light predicts sweet water," "lever pressing produces light," and "sweet water is poisonous." This model would predict reduced lever pressing following devaluation.

In conclusion, we believe that more fully understanding the role of stimulus values can go a long way toward understanding whether revaluation effects are due to chaining or to mental mechanisms that can recombine information.

7.4 Latent Learning

Before revaluation experiments became popular, investigations of thinking-like processes in animals often focused on latent learning of causal information and on insight (the sudden appearance of solutions to problems). This section explores latent learning; insight is covered in section 7.7.1.

Latent learning refers to learning that is not immediately reflected in behavior, and typically that occurs regardless of overt reinforcement (Blodgett 1929, MacCorquodale and Meehl 1954, Bitterman et al. 1979, Rashotte 1979). Traditionally, latent learning has been studied by running rats in mazes, looking for evidence that the animals could learn to navigate the maze simply by exploring it in the absence of reinforcement. MacCorquodale and Meehl (1954) reviewed 48 experiments in mazes and concluded that 30 showed evidence of latent learning. Similar observations also exist in naturalistic settings. For example, mice that are allowed to explore an environment before having to escape from an owl survive better than mice that enter the environment at the same time as the owl (Metzgar 1967).

The classic debate around latent learning concerned two theoretical positions. The "stimulus-response" position is that learning consists of modifying stimulus-response associations and that it occurs only when a response is reinforced. The "cognitive" position holds that learning can occur without reinforcement and can result in flexible knowledge that can be used in planning and reasoning. According to this view, a rat in a maze would learn cause-effect relationships, such as "turning left at point A leads to point B," and would be able to use this knowledge to plan routes in the maze. The stimulus-response position is widely attributed to Thorndike, Guthrie, Hull, and Skinner, and the cognitive one to Tolman (the views of these authors, however, are more nuanced; see Spence 1950, Rashotte 1979, Holland 2008). Eventually, this debate died out as both associative theories (Hull 1952, Spence 1950) and cognitive ones (Tolman 1933,1948) appeared compatible with most findings, though not all (Thistlethwaite 1951, MacCorquodale and Meehl 1954, Rashotte 1979). However, the controversy is often remembered as showing that maze experiments support cognitive theories (Jensen 2006), possibly because associative accounts required lengthy analyses of stimulus-response sequences and were couched in frankly uninviting language (Hull 1943, MacCorquodale and Meehl 1954). Here we hope to show in a more accessible way that associative learning is a viable, and possibly superior, account of latent learning. We first consider in general how latent learning can arise in associative learning. We then look at the classic maze experiments by Blodgett (1929) and Tolman and Honzik (1930b)—which continue to be widely discussed—and other experiments with mazes.

7.4.1 How Associative Learning Can Produce Latent Learning

Latent learning rose to prominence in theoretical debates because it was believed to be incompatible with associative learning. This, however, is not true in general. In A-learning, for example, there are several ways in which learning can occur without overt reinforcement, without overt behavioral changes, or both.

The description of latent learning occurring "without reinforcement" refers to reinforcers such as food, water, or shock that are provided by experimenters. It has been long recognized, however, that animals may perceive other events as reinforcing. In A-learning, these events would have a nonzero u value. For example, being reunited with cage mates after running a maze may be rewarding to a rat (Angermeier 1960, Salazar 1968, Rashotte 1979), while running into the blind alley of a maze and having to backtrack may be punishing (Reynolds 1945, Berlyne and Slater 1957, Jensen 2006). These covert reinforcers may result in latent learning through changes to stimulus values (w values). As we have discussed above in the case of revaluation experiments, these changes are behaviorally silent because w values do not enter decision making (equation 3.8), yet they can change behavior later on by affecting stimulus-response values (v values). For example, suppose a rat experiences a reward when it is picked up from the "goal box" G of a maze (the location where food will later be placed). This reward may derive from being returned to the home cage, from a social bond with a familiar experimenter, or from the maze being somewhat aversive. The important point is that G is experienced before a reward, and so it can gain stimulus value $w(G) > 0$. When the rat is placed in the maze again, $w(G)$ can reward the approach to G and can sum with the value of a food reward at G. As a consequence, rats with the experience of being picked up from G would learn to go to G faster than rats without this experience. In a real experiment, it would be tempting to conclude that the rats had learned "where G is," but an associative analysis shows that this is not the only possible explanation.

A further potential source of latent learning is memory traces. As discussed in section 5.4, animal working memory can typically retain information about past stimuli and responses for up to several minutes. These memory traces may be thought of as faded versions of the original stimuli that can take part in associative learning and decision making in a variety of ways (Amundson and Miller 2008, Fraser and Holland 2019, Ghirlanda et al. 2017). For example, consider a sequence of stimuli

$$S_1 \to S_2 \to S_3$$

(we omit the intervening responses for simplicity). Let a lowercase letter indicate the memory trace of a stimulus that is no longer present, obtaining

$$S_1 \to s_1 S_2 \to s_2 S_3$$

This introduces the possibility that events that change the value of S_i also affect the value of S_{i-1}. In studies of latent learning, and in the revaluation

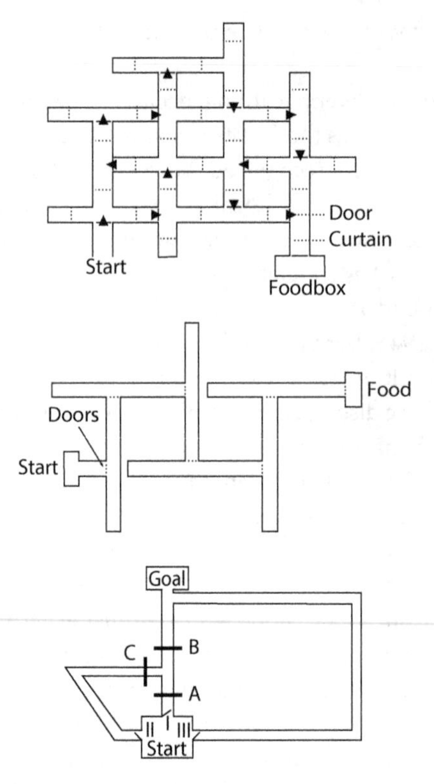

FIGURE 7.3. Famous mazes in animal psychology. Top and middle: Mazes used to study latent learning by Tolman and Honzik (1930b) and by Blodgett (1929). Bottom: Maze used by Tolman and Honzik (1930a) to study insight.

studies considered earlier, memory traces may thus allow learning about a stimulus that is no longer physically present. The role of memory traces in latent learning and revaluation remains to be explored.

In addition to the possibilities just mentioned, we also point out that a "latent learning" experiment may actually feature changes in overt behavior. For example, there can be changes to which parts of a maze are visited even as a result of running the maze in the absence of rewards. As we discuss in the next section, these changes can also be important in understanding what latent learning experiments really show.

7.4.2 Experiments by Blodgett and by Tolman and Honzik

The influential experiments by Blodgett (1929) and Tolman and Honzik (1930b) used similar mazes (figure 7.3) and similar procedures. Blodgett compared a control group of rats, for which reaching the goal box was always

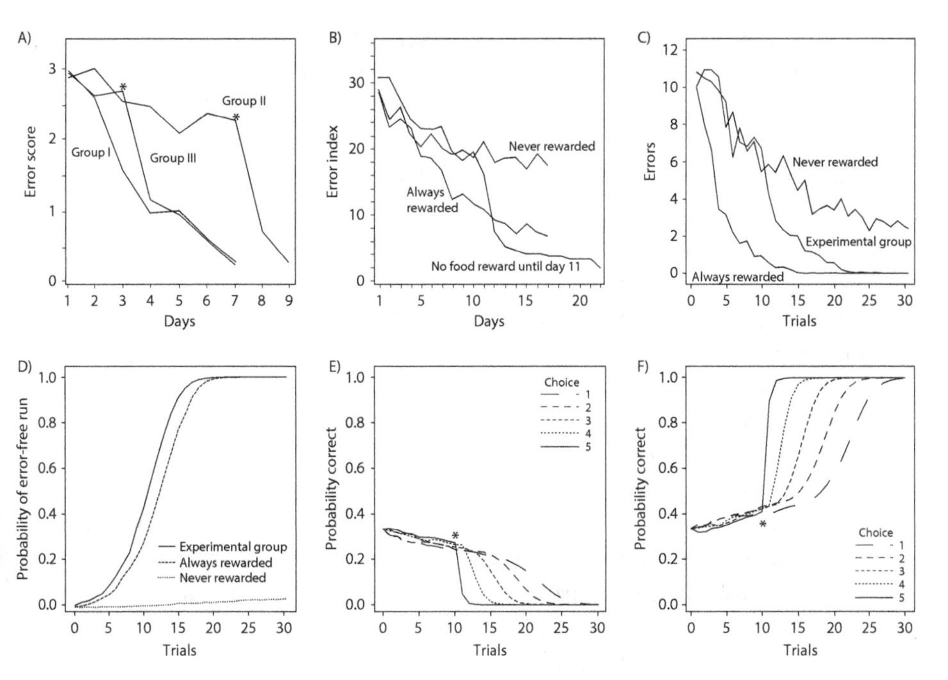

FIGURE 7.4. Latent learning in mazes. A, B: Results by Blodgett (1929) and Tolman and Honzik (1930b) on latent learning in mazes (redrawn from the original sources). In Blodgett's study (A), rats in group I received rewards from day 1, group II from day 4, and group III from day 7 (stars indicate last nonrewarded day for groups II and III). Tolman and Honzik (B) rewarded one group from day 1 and one group from day 11, while a third group was never rewarded. See figure 7.3 for maze plans. C: Simulation of A-learning in a maze with 5 choice points. The experimental group is rewarded from trial 10. Errors are average visits to blinds over trials. D: Probability of an error-free run. The curve for the experimental group starts with the first rewarded trial. E, F: Probability of wrong and correct turns at the five choice points, in the experimental group. Unrewarded trials depress wrong turns in favor of correct turns approximately uniformly across the maze. Rewarded trials show a backward pattern with earlier improvement for choice points closer to the goal. Choice points and blinds are numbered from entry to goal box. See section 1.5.5 for simulation details.

rewarded with food, with two experimental groups, which could explore the empty maze for either 4 or 7 days before food was introduced. Tolman and Honzik used two control groups, one always rewarded and one never rewarded, and one experimental group, for which food was introduced after 11 days. The crucial result of both experiments is that, after food was introduced, the experimental groups learned faster than the control groups (figure 7.4A, B). This led the authors to conclude that the rats had learned a mental map of the maze during the initial days of unrewarded exploration—an example of latent learning that was deemed incompatible with associative learning. Jensen (2006), however, pointed out that the experimental groups improved even before the food was introduced, and that even rats that were never rewarded improved (figure 7.4A, B). Jensen argued that the rats were, in fact, learning to avoid blind alleys, which may be slightly aversive (Reynolds 1945, Berlyne and Slater 1957). This resulted in "improvement" because, due to a quirk of the maze layout, avoiding blind alleys led the rats to travel the route that would later be rewarded (figure 7.3). Thus, when the reward was introduced, the experimental groups were already traveling a mostly correct route, which led them to learn faster than the control groups. This is an example of how a "latent learning" outcome may result from an overt, but unappreciated change in behavior.

We have tested Jensen's account by running the A-learning model in a maze consisting of five connected T-mazes (intermediate in size between Blodgett's and Tolman and Honzik's) and assuming a slightly negative u value for blind alleys. We replicated the design in Tolman and Honzik (1930b). In each trial, simulated rats could move back and forth freely in the maze, until they arrived at the goal box. At this point, they were returned to the start of the maze for a new trial. Two control groups were either always rewarded or never rewarded upon reaching the goal box, while the experimental group was rewarded after the tenth trial. The results of these simulations replicate both the observed improvement during nonrewarded trials (figure 7.4C) and the faster learning in the experimental group (figure 7.4D), shown as the probability of an error-free run starting from the first trial with a reward. Panels E and F provide insight into how this effect arises. Panel E shows that the probability of entering the blind alleys decreases somewhat even during unrewarded trials. At the same time, panel F shows that the probability of making a correct choice increases somewhat. In keeping with Jensen's proposal, these changes make it easier to learn the route to the reward once it is introduced.

According to figure 7.4E, F, unrewarded exploration of the maze affects blind alleys and correct choices approximately equally, regardless of distance from the goal. Once the reward is introduced, however, the pattern is strikingly different: the maze is learned backward from the goal. As seen in chapter 3, this backward pattern is a hallmark of chaining and depends on the fact that, in order to reinforce the correct choices, the value of the goal must first propagate backward in the maze, which occurs by increasing the w value of the stimuli that follow correct choices. Crucially, a thinking mechanism that uses a mental map of the maze would show a different pattern.

Because the mental map is built simply by running the maze, once the reward is introduced the mechanism would be able at once to plan a route to the goal, thus turning the right way at all choice points regardless of distance to the goal (see chapter 11 for details). This difference between associative and cognitive accounts was realized early on, and it holds both for the experimental and control groups (it does not arise from latent learning). In support of associative theory, several experimenters provided evidence that, indeed, rats learn mazes backward from the goal (Spence 1932, de Montpellier 1933, Spence and Shipley 1934).

Overall, the results by Blodgett (1929) and Tolman and Honzik (1930b) appear to support associative over cognitive accounts of maze learning. We have reproduced these results also with somewhat different assumptions. For example, it is not necessary to assume that the blind alleys are aversive. One can also assume that each movement carries a cost, such that it pays to reach the goal box as soon as possible to terminate the trial, or that the end of a trial yields a small reward even on "unrewarded" trials, as mentioned in section 7.4.1. Our simulations do not reproduce one result from Tolman and Honzik (1930b): that the experimental group eventually becomes better than the always-rewarded control group (figure 7.4B). This, however, does not seem a robust result (it was not found by Blodgett 1929; see figure 7.4A). Furthermore, it does not support a cognitive account, which predicts that both groups should eventually have identical knowledge of the maze.

7.4.3 Interim Conclusions about Latent Learning

The preceding section shows that, in the end, the famous experiments by Blodgett (1929) and by Tolman and Honzik (1930a) suggest more the operation of associative learning than of a thinking mechanism. These, however, are just two of many experiments. For example, Seward (1949) reported that rats apparently acquired knowledge of a T-maze by exploring it in the absence of food. On the other hand, Spence and Lippitt (1946) and Kendler (1947) found that sated rats would not learn the location of a food they encountered but did not eat, which they interpreted as supporting the idea that reinforcement is necessary for learning. While we may be inclined to dismiss this literature as "old," the experiments were often carefully controlled and up to rigorous standards. For example, Spence and Lippitt controlled for each rat's individual preferences to turn left or right, and made sure each rat did travel to the food location, approached, and sniffed the food. Seward included three different control groups, testing a total of 135 rats. Overall, the results of latent learning studies were varied and sometimes contrasting. Eventually, contemporary scholars concluded that neither the associative nor the cognitive theories of the time could account for all data (Miller 1935, Spence 1950, Thistlethwaite 1951, MacCorquodale and Meehl 1954, Rashotte 1979).

Our own appraisal of this body of evidence is that stimulus-response associations alone cannot explain latent learning, but also that there is no strong evidence against a more complete associative account that includes stimulus values and chaining. This is especially true since several processes that can support latent learning in chaining remain to be studied formally, such as memory traces, perceptual learning, and motivational processes. Pending these developments, however, we also believe that some findings are particularly difficult to explain as the outcome of a thinking mechanism. Examples are that mazes are learned backward from the goal (Spence 1932, Spence and Shipley 1934), and that repeatedly finding a familiar food at a familiar location when sated is not sufficient to go to that location when hungry (Spence and Lippitt 1946).

7.5 Route Planning and Cognitive Maps

Another classic use of mazes is to investigate whether animals can plan routes using a "cognitive map" of their environment (Tolman 1948, Gallistel 1990, Reznikova 2007). This issue is related to latent learning, as the map would be learned independent of reinforcement, but the focus is on how information is used rather than how it is acquired. A classic experiment uses the bottom maze in figure 7.3 (Hsiao 1929, Tolman and Honzik 1930a). The maze has three routes between a start location and a goal location containing food. After running the maze many times, rats come to prefer the routes according to their length, with I the most and III the least preferred. Once these preferences are established, a block is placed at either location A, which blocks route I, or location B, which blocks both routes I and II. The question is which of routes II and III the animal will choose upon running into each of the blocks for the first time. A simple associative account would predict that animals will always prefer route II regardless of whether the block is at A or B, because encountering the block would not by itself change the established preferences. However, if rats have a mental map of the maze, they would realize that choosing route II is unproductive when the block is at B, and would choose route III in this circumstance.

Most psychology textbooks state that rats do demonstrate such planning ability (Jensen 2006), but the research literature is divided. Anthony (1959) reviewed 13 experiments using this maze and concluded that most claimed demonstrations of planning are amenable to simpler explanations. Anthony noted, for example, that a rat retracing its way from the block in B would first encounter the upper alley of route II, which in some experiments was blocked during the test by a gate at C. Thus an attempt to go right would be frustrated, leading to some extinction of right-turning that would favor turning left (toward route III) at the choice point between routes II and III. Additionally, rats tend to alternate spontaneously between left and right turns (Deacon and Rawlins 2006), so that a frustrated attempt to turn right would also favor turning left at the next choice. The study by Evans (1936), in which 23 of 29 rats chose route II, was the only positive result that Anthony deemed methodologically sound.

Without going into further detail, we note that the results of the studies reviewed by Anthony (1959) range between 0 and 90% of rats choosing route II, with an overall mean of 51%—which is not different from random choice (90 rats out of 176 choosing route II, $p = 0.82$, two-tailed binomial test). Because these studies involved only 7–29 subjects (median of 10), some positive results may derive from chance. Even if we accept that rats can plan in this maze, we must conclude that they do so very unreliably.

More recent analyses of maze studies have reached similar conclusions (Ciancia 1991, Bennett 1996, Jensen 2006, Reznikova 2007), and other early results have also been reconsidered. For example, Young et al. (1967) observed that, in another maze introduced by Tolman and Honzik (1930a), the shortcut behavior consisted of choosing a right turn, which was also the last response performed before reaching the food. After replicating Tolman and Honzik's positive results, Young et al. changed the maze slightly so that the last response was a left turn. In this condition, the rats showed no preference for the shortest route.

Spatial navigation has also been studied in other species, including pigeons and primates, both in the laboratory and in nature (Gallistel 1990, Tomasello and Call 1997, Reznikova 2007, Janson 2014, 2016, Bouchekioua, Blaisdell, et al. 2021, Bouchekioua, Kosaki, et al. 2021). Here, too, many positive results can be explained without appealing to planning. For example, data on navigation between multiple food locations do not provide strong evidence that primates regularly plan their routes. For example, Menzel (1973b) had chimpanzees accompany a human, who hid food at various locations in a familiar enclosure and then released the chimpanzees to observe what route they would take. Although the chimpanzees took shortcuts rather than retracing the route taken when the human was hiding the food, they might have just approached familiar landmarks near the food locations (Bennett 1996). Other research suggests that wild primates do not plan their foraging routes, but rather tend to approach the food source that, among those known, has the best cost-benefit trade-off in terms of distance and food quality (Altmann 1974, Janson 2014). This strategy is compatible with value estimation through associative learning. More generally, associative learning models have been able to replicate many behaviors observed in navigation experiments (Reid and Staddon 1997, 1998, Voicu and Schmajuk 2002, Reid and Reid 2005). Finally, Blaisdell, et al. (2021) and Bouchekioua, Kosaki, et al. (2021) have presented an associative account of novel route taking in pigeons that, like our chaining model, hinges on conditioned reinforcement.

7.6 Sensory Preconditioning

Sensory preconditioning has been brought as evidence that animals can learn causal relationships between nonreinforcing events. In a sensory preconditioning experiment, two stimuli, *A* and *B*, are presented repeatedly, either simultaneously or sequentially in succession, in the absence of reinforcement. This is the preconditioning phase. In the successive conditioning phase, a response is trained to *B* by either Pavlovian or instrumental conditioning. The intriguing finding is that, in a final test following conditioning, *A* can elicit the response trained to *B*. Our primary interest in this finding is that it can be interpreted as the result of learning of causal inference and thinking (Gershman et al. 2010). In the simultaneous case, the information learned would be roughly equivalent to "*A* and *B* are often together, so they predict similar consequences." In the sequential case, it might be "*A* predicts *B*, hence it predicts, eventually, the same consequences as *B*." In both cases, the animal would also be able to use this information to draw the conclusion that the response learned to *B* may also be appropriate to *A*. However, sensory preconditioning can also be interpreted as deriving from perceptual learning and associative learning. In section 5.2, we mentioned that perceptual learning can make stimuli easier to discriminate, but it can also lead to generalization between stimuli that have been often experienced together (McLaren and Mackintosh 2000, 2002, Enquist and Ghirlanda 2005). This would explain sensory preconditioning with simultaneous *A* and *B*. The case with *A* preceding *B* can be explained in the same way, relying on a memory trace of *A* still being active when *B* is presented (section 5.4.2). This interpretation appears compatible with sensory preconditioning results that are not immediately anticipated from causal reasoning, such as that the simultaneous condition is more effective than the sequential one (Rescorla 1980) and that a backward $B \rightarrow A$ sequence during preconditioning can be as effective as a forward $A \rightarrow B$ sequence (Ward-Robinson and Hall 1996, 1998). Given available knowledge, we do not find a causal reasoning interpretation of sensory preconditioning more compelling than one in terms of perceptual and associative learning.

7.7 Fast and Slow Decision Making

One characteristic of human thinking is that it can take considerable time, because it often requires a long series of mental operations (Vygotsky 1962, Anderson 1982, VanLehn 1996). More complex tasks require more time, and performance often increases if people are allowed more time. For example, when given enough time, third-graders can perform similarly to high school students on some cognitive tests (Pezdek and Miceli 1982; see also Ward and Allport 1997, Chabris and Hearst 2003). Evidence of time-consuming decision making in animals would therefore hint that animals also perform sequences of mental operations. A good test of whether animals "spend time thinking," however, would need to control for other reasons why decision making takes time. For example, when deciding between two actions, an animal may hesitate simply due to uncertainty (in associative terms, because the underlying stimulus-response strengths are almost equal). Here we consider two kinds of experiments: classic "insight" tasks and newer paradigms in which animals are given varying amounts of time before responding.

7.7.1 Insight

The study of insight learning was sparked by Köhler's (1924) work on problem solving, in which he concluded that chimpanzees can recombine information from different experiences through an internal, time-consuming process leading to a sudden discovery, similar to the human "a-ha!" experience. His most famous experiments saw chimpanzees discovering that they could climb on boxes or use sticks to retrieve bananas out of direct reach. Thorpe (1963, 110) provided the classical definition of insight "as the sudden production of a new adaptive response not arrived at by trial behavior or as the solution of a problem by the sudden adaptive reorganization of experience." Since Köhler's pioneering experiments, many studies have demonstrated fast acquisition of problem-solving behavior. The field has been effectively summarized by Shettleworth (2012), who concludes that "even if it is possible to identify a distinct class of insightful behaviors, the question remains whether they reflect a distinct cognitive process, *insight*." In fact, contrary to what Thorpe's definition suggests, it is not possible to infer the internal reorganization of information based solely on the speed of learning. For example, problem-solving behavior can arise quickly by stimulus generalization from past experience, because of genetic predispositions, or by combining previously learned behaviors (Taylor et al. 2012, Seed and Boogert 2013).

Epstein and coworkers' studies of problem solving in pigeons provide several illustrations of how insightful behavior can arise from learning mechanisms that do not explicitly recombine information (Epstein 1981, 1985, 1987, Epstein and Medalie 1983). For example, in a replication of Köhler's experiments, pigeons were separately trained to peck at a plastic banana, to climb on a box, and to push a box toward a green spot (Epstein et al. 1984). When confronted with the novel situation consisting of a box and a banana suspended out of reach, the pigeons first hesitated, then attempted to reach the banana and turned alternately toward the banana and the box. After a few minutes, they pushed the box under the banana, climbed on it, and pecked the banana. The pigeons' initial hesitation and subsequent swift solution closely mirror traditional demonstrations of insight, but Epstein et al. explained them in terms of behavioral tendencies arising from previous training. Pigeons initially hesitated because of the presence of stimuli associated with different behaviors (banana with pecking, box with climbing and pushing), such that no behavior was preponderant. Pecking and climbing may then have lost strength during the course of the test, for example, through unsuccessful attempts, at which point pushing emerged. The correct direction of pushing may have been determined by generalization, based on the fact that both the green spot (the trained target of pushing) and the banana (the new target) had been associated with food (Honey and Hall 1989, Bonardi et al. 1993). Once the box was under the banana, the behavioral tendencies to climb the box and peck the banana could sum, leading first to climbing and then to pecking. To validate their analysis, the authors confirmed that training on directed pushing, climbing, and pecking were all necessary for insightful behavior. For example, pigeons trained to push without a target continued to push aimlessly during the test. Overall, this analysis appears compatible with a chaining model of associative learning, and indeed Epstein (a student of Skinner) coined the term "automatic chaining" to refer to the spontaneous emergence of a behavioral sequence based on training single steps of the sequence. Even if an associative explanation is rejected (e.g., Ellen and Pate 1986), this work highlights the importance of previous experience, such that insightful behavior in subjects with an unknown history is difficult to interpret unambiguously.

Work on another problem devised by Köhler, the obstruction problem, leads to similar conclusions. In this problem, access to food is blocked by an obstacle. The insight to be had is to push the obstacle away. However, only two out of seven chimpanzees in Köhler's experiment did so. In later experiments, reviewed by Nakajima and Sato (1993), a chimpanzee, a gorilla, dogs, and cats also failed, while orangutans, baboons, several macaques and crows, and a fish succeeded. Nakajima and Sato speculated that the unsuccessful animals lacked appropriate experiences, rather than being less insightful. To test this hypothesis, they trained pigeons to peck a key for food and then placed a box in front of the key, making it inaccessible yet still visible. The pigeons did not push the block away. However, they did so after having been trained, separately, to push the box around. Like Shettleworth (2012), Nakajima and Sato concluded that insightful behavior does not necessarily imply the specific mental process of "insight."

Shettleworth points out two further observations at odds with the traditional belief that insightful behavior cannot derive from associative learning. First, although insight is typically defined as sudden (Köhler 1924, Thorpe 1963), many studies report gradual behavioral change. Second, associative learning can also result in sudden behavioral changes, even if the underlying learning processes are stepwise and gradual (see section 3.8.4 for an example). In summary, understanding whether observations of insightful behavior arise from associative learning or more complex mechanisms requires focusing on recombination of information rather than on learning speed. Experiments should carefully control the subjects' training history and include groups with different experiences, as in Epstein et al.'s (1984) and Nakajima and Sato's (1993) experiments. Carefully controlled experiments can be illuminating, such as Cook and Fowler's (2014) demonstration that pigeons trained as in Epstein et al. pushed a box *on which they could not stand* as much as a functional box. It is also important to try to reproduce findings of insightful behavior through simulations of associative learning and other mental mechanisms (Lind 2018). Finally, since the concept of insight highlights the importance of internal processing, it may be productive to conduct experiments in which the time available for such processing is manipulated, as considered in the next section.

7.7.2 Time-Consuming Decision Making

We are aware of two studies that have correlated accuracy of decision making with the time allowed to reach a decision. Dunbar et al. (2005) let chimpanzees, orangutans, and children observe puzzle boxes for varying amounts of time before attempting to open them, up to 48 hours in some tests. While children benefited from observing the boxes, the apes did not, leading the authors to conclude that apes cannot perform mental rehearsal. The results, however, may have been confounded by the fact that different puzzle boxes were used in different tests, making it impossible to separate neatly the effects of observation time, practice, and puzzle difficulty (Dunbar et al. 2005).

In another experiment, Lind et al. (2017) subjected Sumatran orangutans to a series of two-choice problems. The apes could choose between two objects: a functional tool that would result in a reward (peanuts or fruit juice) and a dysfunctional object. For example, the orangutans could choose between an intact rake that could be used to reach a peanut and a rake that had been cut in two halves. In one condition, the orangutans were forced to make the choice immediately when presented with the two tools; in the other condition, the orangutans were allowed to observe the two objects for 45 seconds before making a decision. Observing the problem did not increase performance.

7.8 Animal Cognition

As discussed in section 2.2.2, present-day studies of animal cognition often look at animal information processing in terms of planning, reasoning, and similar concepts, which are often contrasted with associative learning. Results from this field confirm and extend earlier findings, such as that animals can learn behavioral sequences and that they have well-developed memory skills (Clayton and Dickinson 1998, Babb and Crystal 2006, Clayton et al. 2007, Inoue and Matsuzawa 2007, Crystal 2010). However, we could not identify results that unambiguously indicate abilities beyond associative learning.

A weakness of some animal cognition research is that it contrasts such concepts as planning and cognitive maps with an impoverished view of associative learning—based only on stimulus-response associations and much less powerful than chaining and other modern associative learning models (Dickinson 2012, Ludvig et al. 2017, Sutton and Barto 2018). This often results in uninformative experiments. The work on revaluation and mazes summarized above shows that distinguishing associative learning from thinking-like mechanisms is difficult, which remains true of modern studies (Correia et al. 2007, Raby et al. 2007, Redshaw et al. 2017, Lind 2018). Furthermore, skills such as episodic-like memory (Crystal 2010) and use of social information (Keefner 2016) that are often discussed as "cognitive" are not per se in contrast with associative learning, as discussed in section 5.5.

A second concern is that the complexity of experimental data—such as failed replications of positive findings, nonpublication of negative ones (Clayton et al. 2020, Farrar et al. 2021), and cases in which behavior agrees more with associative learning than with thinking-like mechanisms—is often overlooked. For example, psychology textbooks frequently claim that the maze experiments discussed above demonstrate the use of cognitive maps (Jensen 2006). A recent example concerns experiments in which corvids were trained to retrieve a food morsel floating in a water container by dropping stones in the container in order to raise the water level and bring the morsel within reach of their beaks (Emery and Clayton 2009, Taylor et al. 2009, Jelbert et al. 2014). Early observations of this and similar behaviors prompted claims that the birds understood the cause-effect chain "drop stones → raise water level → get food." Later analysis, however, revealed many hallmarks of associative learning, such as gradual learning curves, and systematic mistakes, like dropping stones equally often in containers filled with water or with sand or that were empty (Ghirlanda and Lind 2017, Hennefield et al. 2018).

In summary, clarifying what the data really show and delimiting theoretically the true scope of associative learning are both necessary steps toward an accurate understanding of animal intelligence. We consider some of these issues in the next section and again at the end of the next chapter.

7.9 Distinguishing between Associative Learning and Thinking

We conclude this chapter by exploring potential avenues to distinguish between associative learning and thinking. This chapter and the previous ones show that providing unequivocal evidence that rules out associative learning is not easy. As remarked at the beginning of the chapter, thinking and associative learning are both optimizing processes and often produce the same end results (chapter 3; Enquist et al. 2016). Thus it is more informative to look at patterns of behavioral change during learning than at the final outcome of learning. However, it can be difficult to design proper control conditions, for example, because memory changes may be silent (latent learning) and because stimulus situations may share elements that are difficult for the experimenter to identify (including memory traces and other internal sources of information). Finally, we often lack the theoretical understanding necessary to derive unique predictions. We do not understand associative learning completely, and there are few formalized models of causal learning and thinking in animals that can be analyzed with sufficient rigor. For example, thinking is generally assumed to perform better than associative learning, without proof (see section 11.3).

Based on the studies reviewed so far, we can suggest a few directions for empirical research. First, it seems worthwhile to investigate how longer behavioral sequences are learned. The reason is that results from tasks that require just one or two behaviors, such as a single T-maze, are often compatible with both associative learning and thinking (section 7.4.2; MacCorquodale and Meehl 1954, Rashotte 1979). Additionally, the difference in learning patterns and speed between associative learning and thinking increases with sequence length (see figure 11.2).

Another suggestion is to explore predictions that are distinctively different for the two hypotheses. We offer some ideas below. The advantage of this method is that it relies on comparing the behavior of animals with the two predictions, rather than on comparing an experimental and a control group. This increases sample size and avoids the need to control statistically for group differences that may arise spuriously, such as from individual differences and from differences in previous experiences. The latter are largely unavoidable in primates and other rare species, such as the New Caledonian crow and the kea.

7.9.1 The Order of Acquisition

Unless special conditions are arranged, associative learning can only learn behavioral sequences backward from the reward (see chapter 3), including when latent learning facilitates acquisition. In contrast, a thinking mechanism would be able to recombine information regardless of the order of experiences (section 2.5). For example, we have previously pointed out that the observation that rats learn mazes backward from the goal agrees with associative learning but not with thinking.

The difference between chaining and thinking is even more striking when there is no possibility of backtracking, that is, when errors bring the animal back to the start of the sequence (see exit patterns in section 3.5). In this case, most experiences will be about the early steps of the sequence, yet chaining will still learn backward. Consider, for example, a sequence of two behaviors:

$$S_1 \to B_1 \to S_2 \to B_2 \to S_{\text{reward}}$$

In a thinking mechanism, the probability of B_1 increases immediately when S_{reward} is reached for the first time, while in chaining reaching S_{reward} only increases the probability of B_2. The probability of B_1 increases only as S_2 gains stimulus value and thus reinforces performing B_1. This means that we expect a thinking mechanism to learn more quickly, and this difference increases with the length of the sequence (see figure 11.2; see also figure 3.4). Longer sequences in which backtracking is not possible are very difficult for chaining to learn.

7.9.2 Patterns of Extinction

Another approach is to manipulate experiences in revaluation experiments (section 7.3) in order to expose the differences between thinking and chaining. For example, suppose animals learn a sequence of three actions to earn food:

$$\textit{String} \to \text{Pull} \to \textit{Lever} \to \text{Press} \to \textit{Food} \to \text{Eat} \to \textit{Nutrients} \qquad (7.3)$$

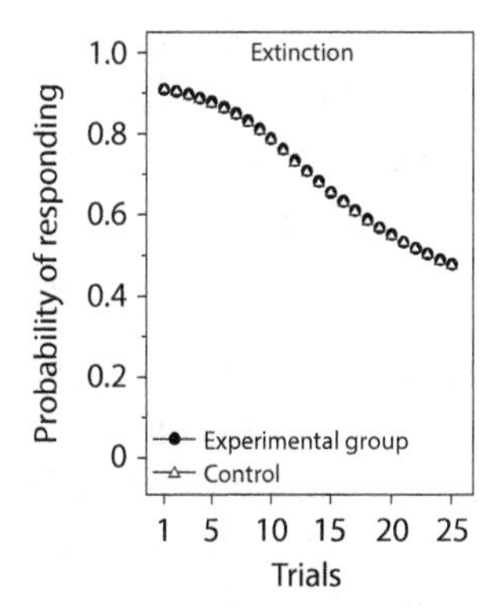

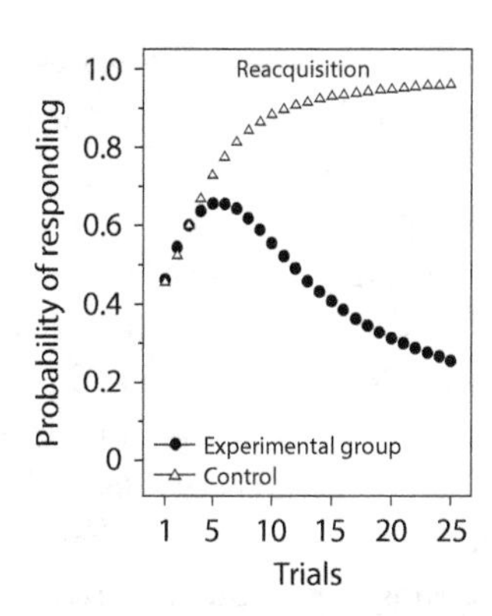

FIGURE 7.5. Learning curves for A-learning, in a revaluation experiment with a sequence of three actions (equation 7.3). Left: During extinction, only the string is available, and pulling has no consequences. Right: During reacquisition, the complete chain is again possible. Learning parameters: $\alpha_v = \alpha_w = 0.1$, $\beta = 1$, $u(\text{food}) = 10$, $u(\text{sick}) = -300$. The cost of pulling and pressing was 1, while not responding was cost-free. Curves show the average over 1000 simulations. Training prior to the test consisted of 100 trials, resulting in a frequency of pulling above 0.9. See section 1.5.5 for simulation details.

Suppose now the food is devalued, so that $w(\mathit{Food})$ becomes negative; and that pulling the string is extinguished, but in the absence of the lever, so that lever pressing is not possible (figure 7.5, left). If the whole sequence is again made possible, including earning the food, chaining predicts that string pulling should first increase before eventually decreasing (figure 7.5, right). The reason is that it takes a number of experiences for the negative $w(\mathit{Food})$ value to back-propagate along the sequence and eventually decrease $v(\mathit{String} \to \text{Pull})$. In the meantime, string pulling continues to be reinforced by the initially positive $w(\mathit{Lever})$, which was established when learning the sequence. On the other hand, a mental mechanism that can recombine causal information from different experiences (i.e., think) would immediately decrease string pulling, as it would be able to infer directly that string pulling is no longer desirable.

7.9.3 Do Animals Conflate Learned Value and Primary Value?

In chaining, learned stimulus value (w) and primary value (u) are treated as equivalent, because stimulus value is a computational device that enables the algorithm to take into account *future* primary value. In other words, it is only because stimulus value reinforces behavior in the same way as primary value that chaining can take into account the future consequences of a behavior.

Formally, this equivalence is expressed by learning equations 3.11 and 3.12, which contain the sum $u + w$ but not these values separately. Thus different (u, w) pairs, such as $(0, 10)$, $(5, 5)$, and $(-5, 15)$, have the same effect because they all sum to 10. Whether learned and primary values are truly equivalent to animals has been researched in some detail, providing a hitherto unrecognized way to distinguish chaining from thinking.

For example, Zimmerman and Hanford (1966) delivered food to pigeons at random times, regardless of their behavior. Food availability was signaled by several stimuli, such as a light illuminating the food receptacle and the sound of the food delivery mechanism. The pigeons could produce the stimuli by pecking a key, but this never yielded food. The pigeons learned to peck the key nevertheless, and continued doing so until the experiment was terminated several *months* later. This result is readily interpreted in terms of chaining, with the light and sound maintaining stimulus value throughout the experiment, and thereby remaining capable of reinforcing pecking (see Zimmerman 1963, 1969, for similar results). On the other hand, we would expect a thinking mechanism to understand that pecking never leads to food (see the model of thinking in section 11.4). Many similar results appear in the literature on conditioned reinforcement. For example, Williams and Dunn (1991) showed that, of two equally rewarded actions, pigeons prefer the one leading to more presentations of a stimulus with learned value (a light that accompanied food delivery). Further, Clark and Sherman (1970) showed that pigeons' performance on a match-to-sample task (section 5.4) worsened considerably if incorrect responses were followed by a stimulus with learned value. In this case, the pigeons' fondness of learned value not only was unproductive but actually led to losing food. Cronin (1980) obtained similar results in the experiment reviewed in section 3.7. We believe that expanding this line of research could provide stimulating opportunities to distinguish chaining from thinking, for example, by testing animals other than pigeons.

8

The Nature of Animal Intelligence

- We present a standard model of animal intelligence that combines chaining with genetic predispositions that guide learning toward adaptive outcomes.
- The model can account for major observations often deemed beyond the scope of associative learning.
- We remain open to the possibility that, in some species and domains, mechanisms exist that go beyond the standard model.
- Further study of animal intelligence will benefit from formalizing putative mechanisms mathematically and deriving predictions to compare with behavioral data and the standard model.

8.1 Introduction

In this chapter, we attempt to draw some general conclusions about animal intelligence. Building on the three hypotheses about animal intelligence laid out in chapter 2, we suggest the *standard model,* described in section 8.2. This model recognizes the power of associative learning, the costs of behavioral flexibility, and the need for genetic guidance of learning and decision making. We have suggested in the previous chapters that these elements can explain most empirical observations of animal intelligence without assuming that animals can mentally recombine information. The standard model also synthesizes ethology and learning psychology, combining the species specificity of behavioral adaptations with the generality of associative learning. After describing the standard model, we relate it to our broader discussion of thinking in animals.

8.2 The Standard Model

This section describes the standard model from a conceptual point of view. Formally, most aspects of the model can be expressed in terms of the

A-learning model of chaining in equations 3.8, 3.11, and 3.12. We summarized how genes can influence chaining in table 4.2 and what information animals use in chapter 5. Both genetic influences and sources of information can be modified by natural selection to adapt behavior to particular ecological niches.

The standard model aims to specify what associative learning in animals is and how it is integrated with the rest of the behavioral machinery. Another purpose is to help us recognize both the power of associative learning and its limitations. Thus the model can help design experiments that can clearly distinguish between associative learning and other mental mechanisms. We describe the model in three parts: decision making, learning processes, and organization of behavior as a whole.

8.2.1 Decision Making

Decision making selects a behavior by integrating information from many sources, including external stimuli, memory, and internal factors (chapters 3 and 5). Inborn and learned factors guide how much weight is given to different kinds of information (chapter 4).

STIMULI

External stimuli are key determinants of effective behavior. Generally, decision making is not directly based on sensory input, but on elaborations of sensory input that provide more relevant information than raw sensations (chapter 4). Decision making is mainly based on the current stimulus situation. Barring specific adaptations, the ability to make decisions based on stimulus sequences is limited and derives from short-lived memory traces of stimuli (chapter 5).

MEMORY

Memories allow decision making to take past experiences into account. In the standard model, long-term memory consists of stimulus-response values and stimulus values, with only the former entering decision making directly (chapter 3). Although this may seem limiting, the decision-making machinery can use any number of such memories, and genetic evolution can establish memories that serve both general and species-specific needs (chapters 4 and 5). For example, by using both a fast- and a slow-changing value for the same stimulus-response pair, it is possible to simultaneously store both short- and long-term information about the environment (chapter 5). Moreover, specialized memory systems can store information about specific kinds of events (e.g., food locations) without incurring the combinatorial costs arising from remembering numerous experiences. Memories also allow comparing actual outcomes with expected values stored in memory. If there is a substantial mismatch between expected and experienced outcome, the animal can respond with, for instance, aggression or exploration (section 5.3).

INTERNAL FACTORS

A variety of internal factors also influence decision making. Internal stimuli can originate from sense organs within the body that monitor physiological variables, such as energy and water reserves. Internal clocks and motivational and emotional factors can also control and organize behavior. Including these factors in decision making enables both focusing on one particular activity at a time and organizing different activities over longer time spans. While the general machinery for control and organization of behavior is inborn and adapted to the species' niche, experiences can modify the weight given to at least some internal factors in supporting different responses.

BEHAVIORAL REPERTOIRE

The outcome of decision making is a behavior. Most animals have a small inborn and context-dependent repertoire of behaviors, adapted to the species' niche. The reason why the repertoire is small is that large repertoires incur exponentially larger learning costs (section 4.4). Some species, however, have large behavioral repertoires in particular domains, such as vocalizations in songbirds and hand movements in primates (see section 8.3.4 for these exceptions).

EXPLORATION

To learn about the utility of different behaviors, their consequences must be explored. Therefore, decision making in animals typically generates some variation in the selection of behavior. This allows animals to learn about new stimuli, find new solutions to problems, and adapt to a changing world. The basic regulation of exploration is inborn, context-dependent, and species-specific (chapter 4). Exploration can in some species become an activity on its own, such as play behavior that does not fulfill any current goal but presumably prepares the individual for the future (section 4.4). Opportunities for such exploration are greater in species with parental care, since the young do not have to sustain themselves right away.

8.2.2 Learning

Learning processes and memories are shaped by genetic evolution. Genetic predispositions regulate under which conditions learning occurs, at what rate something is learned, and what can be learned (chapter 4). Genetic predispositions also specify values for primary reinforcement, determining what the animal strives for. Learning processes, both general and specialized, have coevolved with the decision-making machinery and are based on associative learning principles (chapter 3). Specifically, in the standard model, animals do not represent information in a way that allows for internal or mental reorganization of information.

Neither overt reinforcement nor immediate behavioral change is necessary for learning in the standard model. Animals may find any stimulus reinforcing, provided that the genes or previous experiences have assigned value to it. Social stimuli are an example, enabling observational learning, such as imprinting (chapter 6). In addition, stimulus values can be learned without changing behavior directly (sections 7.3 and 7.4), and perceptual learning can tune internal representations of events based on experience.

LEARNING ABOUT THE VALUE OF RESPONSES

Learning about the value of responses is accomplished by stimulus-response values, which are akin to stimulus-response associative strengths. Functionally, a stimulus-response value for a stimulus S and a behavioral response B can be interpreted as a subjective estimate of the value of responding with B toward S. Stimulus-response values are typically statistical summaries of many experiences (section 5.3) but can also represent specific circumstances (section 5.4). The value of a stimulus-response pair includes the value of expected future events, as discussed next. This yields the ability to learn productive behavior sequences without an explicit mental representation of the sequence (chapter 3).

LEARNING ABOUT THE VALUE OF THE FUTURE

In addition to learning about the outcome of responses, animals also learn about expected future events, by attaching values to stimuli. These values indicate what can be expected from future stimuli and behavior (chapter 3). They do not influence decision making directly, but rather act as "conditioned" reinforcers that support the acquisition of productive behavior sequences (section 3.4). Productive behavior sequences can be established independently in different behavioral domains (e.g., resting, foraging, drinking, or courtship) by using domain-specific stimulus values, even if some stimuli are shared between domains. For example, the sight of a male may have different values for a female, depending on whether she is looking for food or for a mate. (Section 3.4 discusses how stimulus-response values and stimulus values relate to instrumental and Pavlovian conditioning.)

PERCEPTUAL LEARNING

Perceptual processing is to a large extent inborn, but it is also subject to learning (section 5.2). Such learning occurs without reinforcement to organize the perceptual system to better represent the animal's environment, but also in response to feedback from the environment that highlights some stimulus features as meaningful. Perceptual learning is important to reduce combinatorial problems in perception, such as in enabling animals to categorize collections of sensory inputs as originating from the same object (Goldstone 1998, Dosher and Lu 2017). We have not covered perceptual learning in detail because we are not aware of crucial differences between animals and humans.

SPECIALIZED LEARNING

The general machinery for associative learning can also be used for specialized purposes, with genetic predispositions guiding what is learned, how quickly learning proceeds, and when learning occurs. Examples of these specialized learning systems are imprinting and song learning in birds, which may also include specialized memory templates.

8.2.3 Organization of Behavior

Behavior is also organized at a higher level than immediate decision making. This organization provides focus and endows behavior with goal-seeking properties, as well as ensuring that different needs are fulfilled. For example, to survive an animal must drink and feed. However, these needs require different behaviors and often must be fulfilled at different locations. Although we have not presented a formal model, we envision that different states of need can influence decision making by focusing the animal on certain properties of stimuli. For example, an associative learning system may assign different values for a location—one relevant to foraging, one to escaping predators, one to finding mates, and so on. Which value is relevant for decision making would then be selected by what the animal is seeking, as encoded in its motivational state.

Apart from its daily organization, behavior is also often organized seasonally, over development, and for reproductive purposes. A temporal and hierarchical organization of activities can be controlled by motivational, emotional, and hormonal states that define the current focus of the behavior system, such as foraging, avoiding danger, or seeking a mate. The behavior system switches between such states as a consequence of both internal and external stimuli, perhaps in a way similar to decision making.

8.3 Explanatory Power

As mentioned several times in the previous chapters, the ability of associative learning to explain animal behavior and intelligence has regularly been questioned. Below, we discuss the extent to which the standard model can provide a complete picture of animal behavior and intelligence.

8.3.1 Efficiency

Whereas some have discarded associative learning as an explanation for animal intelligence (e.g., Byrne 1995, Tomasello and Call 1997, Penn et al. 2008, Sanz et al. 2013, Povinelli 2020, 431), the standard model demonstrates that it is indeed powerful enough to yield intelligent behavior (chapters 3 and 7; see also chapter 11). In chapters 3, 6, and 7, we discussed how associative learning can result in optimal behavior, and how it can account for behavioral phenomena often considered to be beyond the scope of associative learning.

8.3.2 Animal Intelligence Can Be Superior to Human Intelligence

The standard model may also be able to clarify why humanlike thinking is not widespread among animals. If it correctly captures the mechanisms of animal behavior and intelligence, we must ask why genetic evolution has equipped animals with this kind of intelligence, rather than with humanlike intelligence. One possibility is that human intelligence represents a technically complex feat that, so far, has had the time to emerge in only one species. Additionally, there is also the possibility that associative learning is superior to thinking in many circumstances, as mentioned in chapter 2. To the best of our knowledge, this hypothesis has not been seriously considered, as thinking is implicitly considered superior to associative learning. However, we demonstrate that this is not necessarily true in chapter 11.

8.3.3 Goal Seeking, Expectations, and Self-Control

Associative learning has often been criticized for being unable to explain future-oriented behavior, such as when animals strive for a specific goal (McFarland and Bösser 1993) or forfeit an immediate reward for a larger, delayed one (Evans and Westergaard 2006, Raby and Clayton 2009, MacLean et al. 2014). However, as we have seen in section 3.8.3, this critique holds only for simple stimulus-response learning (see also Lind 2018). Once stimulus value learning is added, associative learning becomes capable of optimizing sequences of behavior, eventually behaving as if calculating complex cost-benefit trade-offs. The model we have discussed so far represents only the overall value of goals and does not allow for specific goal-seeking behavior, such as obtaining water or particular nutrients (Blair-West et al. 1968, Rozin 1976, Deutsch et al. 1989). However, such goal seeking can be introduced by augmenting associative learning with a motivational system that makes certain outcomes more valuable than others, based on needs and other factors. Such a system would enable, for example, a thirsty animal to select actions associated with water rather than with food.

A similar point can be made about animals showing expectations of future events, such as responding to the omission of expected food with an extensive search or with aggressive behavior (Roper 1984, Haskell et al. 2000). At first sight, associative learning seems incapable of supporting such behavior, because it does not formulate explicit predictions about forthcoming events (Cheney et al. 1995, Pearce 2008, Bird and Emery 2010). However, our chaining model computes the extent to which unexpected value occurs by taking the difference between the actual and expected values of an action's outcome (see equations 3.11 and 3.12). That these value differences drive learning is a cornerstone of associative learning theory (Kamin 1969, Rescorla and Wagner 1972). It has also been suggested that the differences between actual and expected values influence behavior directly through genetic predispositions (Dollard et al. 1939, McFarland 1971, Mackintosh 1974). The behavioral effects of these predictions about outcomes depend on detailed assumptions of the model, but it is conceivable that many observations of animals reacting to unexpected events may be explained with an associative learning model.

8.3.4 Generality of Learning and Species-Specific Adaptations

Experimental and theoretical work in animal psychology strongly supports the idea that learning processes operate similarly across species and contexts (Macphail and Barlow 1985, Macphail and Bolhuis 2001, Colombo and Scarf 2020), such as in associative learning (Mackintosh 1983a, Roper 1983). This generality seems at odds with another strongly supported finding: that animal behavior is species-specific and adapted by genetic evolution to fit particular niches (Hinde and Stevenson-Hinde 1973, Westneat and Fox 2010). However, it is possible to reconcile these findings into a theory that synthesizes ethological and psychological theory (Enquist et al. 2016, Frankenhuis et al. 2019). As presented in chapters 3, 4, and 6, the standard model integrates genetic and learned influences on behavior to explain how animals learn efficient behaviors in their natural environment. The mix of these influences on behavior reflects a trade-off between the potential benefits of learning and its time costs (chapter 4; see also Ghirlanda, Enquist, et al. 2013b, Enquist et al. 2016). The integration of genetic and learned elements can explain why even learned behavior can be species-specific and appear instinctual. For example, dormice, squirrels, and wood mice reliably employ different techniques to open hazelnuts (Yiğit et al. 2001), even though these techniques are partly learned (Eibl Eibesfeldt 1963). According to the standard model, all three species learn through chaining but develop different techniques because of species-specific predispositions and genetically defined behavioral repertoires.

Of relevance to our discussion of animal intelligence, the standard model suggests that species such as primates, crows, and parrots may appear particularly intelligent because their life histories promote more exploration and opportunities for social learning and because they have coadapted predispositions, such as for manipulating objects, rather than because they have evolved more sophisticated mental mechanisms (section 4.5). Genes, in fact, can regulate exploratory tendencies and behavioral repertoires, and an animal that spends more time exploring and using more behaviors is bound to discover more and longer behavioral sequences.

8.3.5 *Lack of Language*

Teaching animals humanlike language has been attempted many times, unsuccessfully (e.g., Terrace et al. 1979, Wynne 2008). Despite many examples of sophisticated communicative skills (Bradbury and Vehrencamp 2011), no animals have been observed using language in the wild. For example, there is no evidence of meaning determined by syntax, such as order of meaning-carrying units (Bolhuis et al. 2018, Fishbein et al. 2020). One potential reason for the poor linguistic abilities of animals was put forward in chapter 5, where we suggested that animals cannot faithfully represent and process stimulus sequences—a likely prerequisite for learning and producing humanlike language. Although many experiments have trained animals to discriminate between sets of stimulus sequences (Gentner et al. 2006, Herbranson and Shimp 2008, Chen et al. 2015, Suzuki et al. 2016), the resulting performance can be reproduced with high accuracy by a simple trace model of sequence memory (Ghirlanda et al. 2017). Because a trace memory can distinguish between many sequences, the sheer observation that animals can solve some sequence discriminations does not imply a rudimentary language ability, especially as the trace model predicts the systematic errors found in sequence discrimination studies (Ghirlanda et al. 2017, Bolhuis et al. 2018).

8.3.6 Culture in Animals

The standard model is consistent with animals possessing behavioral traditions and simple forms of culture, as observed in, for example, fish (Helfman and Schultz 1984, Warner 1988), primates (Whiten et al. 1999, Perry and Manson 2003, van Schaik et al. 2003), birds (Catchpole and Slater 2003, Aplin 2019), and whales (Rendell and Whitehead 2001; but see Mercado III and Perazio 2021). The low complexity of these cultures suggests that they can be entirely the product of associative learning combined with genetic predispositions, according to the mechanisms reviewed in chapter 6 (see also Heyes 2012b, Lind et al. 2019).

We are not aware of cumulative culture in animals, that is, cultural traits that are beyond what a single individual could invent in a lifetime (Dean et al. 2012, 2014, Aplin 2019, Bandini and Harrison 2020). Cumulative culture requires efficient transmission between generations, as young individuals must become proficient with existing cultural behaviors before they can improve on them. However, experimental and theoretical work suggests that simple cultures (Whiten et al. 1999, van Schaik et al. 2003), such as those seen in wild great apes, can persist with low-fidelity cultural transmission. For example, Acerbi et al. (2020) proposed that cultural transmission in animals is often indirect, consisting mainly in younger individuals making use of opportunities created by experienced individuals. For instance, a younger animal can learn about a particular food as it follows an adult around. When exposed to the food, the young animal can develop skills to exploit that resource through individual learning. Supporting this suggestion, individual apes have been shown to reinvent rather readily a variety of tool use behavior seen in the wild, if provided with adequate materials (Bandini and Harrison 2020, Reindl et al. 2018). This has been corroborated further by experimental work on more than 50 chimpanzees, suggesting that the presence of cultural behaviors in a group facilitates the acquisition of those behaviors by individuals, but also that chimpanzee culture is not cumulative beyond what can be achieved through individual learning (Vale et al. 2020). In our analysis, circumstances that favor such learning amount to favorable entry patterns to sequence learning problems (section 3.5). They can facilitate learning but do not create anything much beyond what a single individual can accomplish.

8.3.7 Emotions and Inner Life

Behavioral observations, physiological studies, and neuroscience show that many animals share emotional and motivational systems with humans, including systems for pain, fear, hunger, love, and so on (Panksepp 1998,

Panksepp 2005, Bekoff 2008, Braithwaite 2010). Associative learning has sometimes been portrayed as implying that animals are robots devoid of feelings and inner life, but nothing in associative learning models requires this conclusion. In fact, the standard model suggests that emotions can be an integral part of the mental machinery of animals, through motivational systems, inborn predispositions, and attribution of value to stimuli. The latter is particularly interesting because it suggests that animals (and humans) can learn to connect emotions to stimuli that initially were neutral to them.

8.4 Thinking in Animals?

In the previous section, we suggested that associative learning may offer a complete explanation of animal intelligence, without denying that animals can work for future goals and experience emotions. But is it true that animals have no mental mechanisms that could qualify as thinking? When we started this project, we were confident to find at least some evidence that animals can think. After all, our definition of thinking as the recombination of information, presented in section 2.5, is consistent with those used by biologists and psychologists, such as Thorpe's (1963) definition of insight (section 7.7.1) and definitions of planning and causal reasoning (Raby et al. 2007, Correia et al. 2007, Cheke and Clayton 2012, Clayton et al. 2003, Mulcahy and Call 2006). As we progressed in our work, we were surprised to find that much of the evidence brought in support of animal thinking was also compatible with associative learning, and sometimes favored it. Overall, we found no compelling evidence that animals display any of the three features of thinking discussed in chapter 7: generalized gathering of causal information, recombination of causal information, and time-consuming decision making. Rare observations of immediate changes in response following a single experience come closest to showing that animals can recombine causal information, but even these might arise from associative learning (section 7.3.2). Observations of animal "planning" and "reasoning" have also been explained as the outcome of associative learning (Suddendorf et al. 2009, Kutlu and Schmajuk 2012, Ghirlanda and Lind 2017, Redshaw et al. 2017, Bowers and Timberlake 2018, Hennefield et al. 2018, Lind 2018, Hampton 2019). In a comprehensive review, Penn et al. (2008) reached a similar conclusion, recognizing a gap in mental capacities between humans and animals, although they suggested that associative learning cannot explain animal intelligence completely.

Even though we have reviewed the literature extensively, we cannot be sure that animals cannot think. Thinking may be rare, it may take forms we have not considered, it may be difficult to distinguish from chaining (though see section 7.9), or we may have missed evidence. Nevertheless, our analysis indicates that we should not discard, as yet, the possibility that associative learning may account completely for animal intelligence. Other researchers have expressed similar views (Macphail 1982, Macphail and Barlow 1985, Shettleworth 2010a, Dickinson 2012, Heyes 2012a, Haselgrove 2016, Colombo and Scarf 2020). Many animal cognition researchers will argue that our conclusion grossly underestimates animal intelligence, while some experimental psychologists may think it self-evident. These opinions can coexist because our theoretical understanding is incomplete. Associative learning is often misunderstood (see the previous section and section 2.2.2), and alternative mechanisms are typically described verbally rather than formally, making it difficult to test them with data (Haselgrove 2016). For example, to interpret the revaluation experiments in section 7.3.2, we had to drill down to the level of single experiences in our analysis. This was possible because we had a well-defined mathematical model and a simulation environment, but relatively few animal experiments have been assessed with this level of detail.

A further difficulty in evaluating animal intelligence is that we understand little of the evolution of learning mechanisms. From an evolutionary point of view, thinking matters little unless it impacts fitness. It is tempting to assume that thinking is always superior to associative learning, but this assumption neglects the costs of thinking, such as higher requirements for memory storage and time-consuming information gathering and decision making. We see in chapters 11 and 16 that thinking is superior to chaining only under specific circumstances. This is consistent with thinking being rare and associative learning widespread.

9

Uniquely Human

- Uniquely human behavior includes extensive collaborative sequences, language, and other sequential activities, such as music and dance.
- These sequences can be very long, last a long time, and fulfill a multitude of goals, sometimes remote from biological goals.
- Humans can store and process sequential and episodic information with greater accuracy than other animals.
- Human sequential behavior is underpinned by a diversity of mental skills that operate on sequential information. These skills are often time consuming.
- Most uniquely human skills are of recent origin and not part of our biological history.
- Uniquely human skills do not develop spontaneously. Specific experiences, training, and social environments are necessary.
- Apes and other animals raised by humans do not acquire human-level abilities.

9.1 Introduction

The human and chimpanzee lineages split 6–8 million years ago, yet the first evidence of uniquely human traits dates to the time of the late *Australopithecus* and early *Homo* species, around 2.5 million years ago. Simple stone tools are the earliest finds, but over time more sophisticated artifacts appeared, including small statues, cave paintings, and evidence for burial rituals. These are the first signs of the human evolutionary transition introduced in chapter 1. Since then, humans have transformed the Earth to such an extent that it has been deemed appropriate to introduce a new geological era, the Anthropocene (Steffen et al. 2006, Barnosky et al. 2011). In terms of biomass, humans plus their livestock and pets comprised 0.1% of all vertebrates just 10,000 years ago, whereas today this figure has grown to 98% (MacCready, in Dennett 2009). These extreme circumstances have frequently led to the conclusion that humans have unique mental abilities, but pinpointing what these abilities are and understanding their evolution remains challenging (Penn et al. 2008).

The overarching theme of our book is that sequential abilities are a helpful focal point. Thus far, we have discussed how animals learn behavioral sequences, and from here on we focus on humans. We begin in section 9.2 with a summary of human sequential abilities, both behavioral and mental. Section 9.3 discusses when these abilities may have appeared, while section 9.4 discusses how they develop during an individual's lifetime. Finally, section 9.5 summarizes the main hypotheses about human cognitive evolution.

The remainder of the book presents our ideas about the human evolutionary transition. In chapter 10, we present the hypotheses introduced in chapter 2 in more detail; in the chapters that follow, we describe their most important consequences in terms of intelligence and thinking (chapter 11), acquisition and transmission of sequential information (chapter 12), transmission of mental skills (chapter 13), and cooperation (chapter 14). We then explore the role of cultural evolution in chapter 15. Finally, in chapter 16, we discuss why the human evolutionary transition has not occurred in other animals.

9.2 Uniquely Human Traits

9.2.1 Behavior

Humans use a staggering diversity of behavioral sequences to achieve a multitude of goals, to an extent that is not observed in any other animal. The sequential structure of human behavior has been discussed repeatedly (e.g., Lashley 1951, Heyes 2001, Terrace 2005). Even seemingly trivial daily activities, such as getting dressed, preparing and eating food, or driving a car, have a complex sequential structure. Crafts like pottery, metallurgy, and tool making require long sequences, lasting hours to weeks (figure 9.1), and pursuits like agriculture and obtaining a college degree require years. Many of these sequences are executed in parallel and require accurate task switching to complete successfully. Humans are also masters of communicative sequences, through language, and of expressive sequences, such as in dance and music. The goals that humans pursue through these sequences are as diverse as the sequences themselves. For instance, humans mass-produce objects by the millions, pursue hobbies like sports and gardening, engage in art shows and musical performances, fulfill social functions through rituals and procedures like weddings and court cases. While the biological function of almost all animal activities is clear, many human activities are remote from the basic biological needs of food, rest, safety, or reproduction.

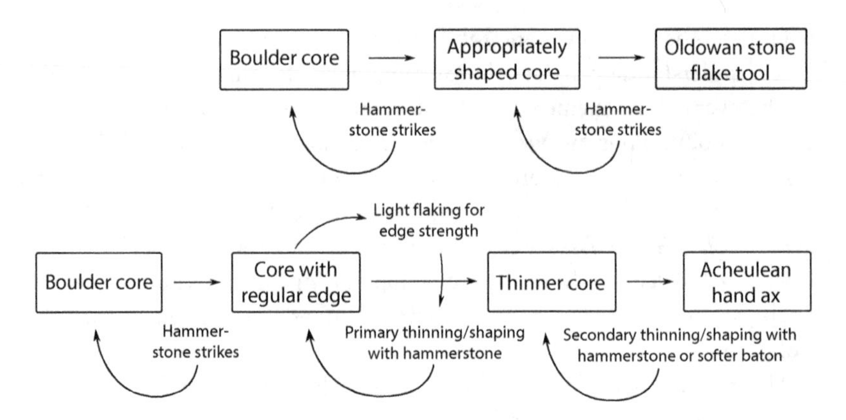

FIGURE 9.1. Coarse descriptions of behavioral sequences used in stone tool production (Stout et al. 2008, Stout 2011). Top: Oldowan stone flake tools. Bottom: Acheulean hand ax.

A major factor contributing to the diversity of human behavioral sequences is our motor flexibility, which is much greater than in most animals (chapter 4). Practically all elements that make up skilled performance are learned, such as phonemes and syllables, piano chords and arpeggios, basketball shots and blocks. The tremendous motor flexibility of human hands is a classic topic in anthropology (Musgrave 1971, Vauclair 1982, Niewoehner et al. 2003), but almost all human movements are highly trainable. Another factor contributing to sequence complexity is cooperation (Carneiro 2003), which we discuss in chapter 14. For example, people may come together to prepare food or build a house, taking advantage of different skills through division of labor. Human cooperative sequences can also involve long production chains in which individuals or groups collaborate while being almost anonymous to each other. For example, consider a blacksmith forging metal axes. Her raw material comes from a long iron production chain that includes mining, processing of iron ore, and transport. Her final product may be sold, for example, to shipbuilders that make boats to sell to fishermen who ultimately produce food. Within these cooperative sequences, language is used extensively for planning and coordination of different parties.

9.2.2 Memory

In chapter 5, we noted that, in animals, there is little evidence for general-purpose episodic memory, that working memory is of comparatively short duration, and that the representation of serial order is inaccurate. Human memory appears superior in all these respects. Human episodic memory enables us to remember and use information from arbitrary events, such as

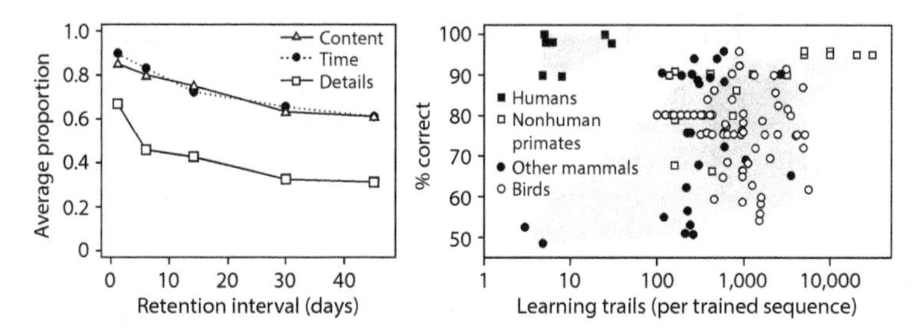

FIGURE 9.2. Left: Retention of episodic information in humans, plotted as average proportion of items retained over time (Kristo et al. 2009). Substantial variation exists in human memory: some events are never forgotten, while others are lost even faster than in the figure. Right: Results from stimulus sequence discrimination studies (Ghirlanda et al. 2017). Humans stand out as attaining better performance with much less training than animals (section 5.4.2).

what happened yesterday at work or on our last birthday (Tulving 1972, 2001). Human semantic memory is also extraordinarily developed and encompasses memory for diverse facts and compositions, like poems or musical scores (figure 9.2, left). We find it particularly significant that memory for sequential information appears much more developed in humans than in animals, as sequential information is prominent in human skills, like language, mathematics, and music (table 9.1). The right panel of figure 9.2 summarizes studies in which animals and humans were trained to discriminate between ordered sequences of stimuli. Human subjects performed better even with stimuli that were more familiar to the animals, such as studies in which finches and humans had to discriminate between sequences of birdsong syllables (Seki et al. 2013, Chen et al. 2015). Additionally, the animals in these studies required 10 to 100 times longer training than the humans (see Ghirlanda et al. 2017, for the methodological details behind making such comparisons). Additionally, Read et al.(2022) conclude that chimpanzee working memory can hold 2 ± 1 items, compared to estimates of 7 ± 2 items in humans (Miller 1956.)

Human memory is not perfect: we forget, misremember, and privilege some information over other (Simons and Chabris 1999, Howes 2006, Lieberman 2011). These facts notwithstanding, it seems safe to conclude that human general-purpose memory is longer term, higher capacity, and more accurate than typical animal memory.

9.2.3 Mental Processes

Humans routinely recombine causal information from different experiences, an ability that we labeled as thinking (section 2.5) and that we found lacking in animals (chapters 7 and 8). These mental processes may be time consuming

Table 9.1. Importance of sequential information in human abilities

Phenomenon	Role of sequential information	Example
Episodic memory	Enables ordering of events	I arrived first ≠ You arrived first
Causal learning	Enables attribution of cause and effect	Noise causes prey to escape ≠ Prey escaping causes noise
Planning	Plans are ordered sequences of actions	Turn door handle, then push ≠ Push door, then turn handle
Imitation	Order of actions must be perceived and remembered	Peel the banana, then eat it ≠ Eat the banana, then peel it
Language	Contributes to meaning	Killer whale ≠ Whale killer
Social intelligence (theory of mind)	Enables attribution of beliefs and knowledge to individuals	You spoke to her before me ≠ You spoke to me before her
Cooperation	Order of actions must be agreed upon and remembered	I count to three, then we lift ≠ We lift, then I count to three
Music	Order determines aesthetic qualities (e.g., melody)	CDEC ≠ CECD
Mathematics	Order of operations influences results	$3 \times 5 - 2 \neq 3 \times (5 - 2)$

Source: Ghirlanda et al. (2017).

and mixed with fast processes, such as associative learning (dual process theory: Sun 2001, Kahneman 2011) For example, many human activities are planned, and this typically occurs through the mental rehearsal of alternative sequences of actions (Gilbert and Wilson 2007). For instance, we regularly plan our daily activities to fulfill one or several short-term goals, like a plan for picking up the children from school, fetching a parcel from the post office, or buying groceries on the way home from work (Miller et al. 1960). The reason why these activities can be labeled as "routine" is that humans are good enough at planning that plans for familiar settings can be devised on the fly almost effortlessly. In reality, routines—in the sense of perfectly repeating activities—are almost nonexistent.

Longer-term examples of planning include stocking up for the winter and planning for an education, a career, or retirement. As a principle for structuring behavior, planning is very different from the genetically guided associative learning that is prevalent in animals. Planning relies on a mental model of the environment, which is explored mentally to find productive sequences of actions. Guided associative learning, on the other hand, selects one behavior at a time based on genetic information and the estimated values of actions (chapter 8). Indeed, part of human mental activity is directed toward learning mental models (understanding how the world works); for example, using logical inference, imagined scenarios ("what would happen if..."), and communication with others either in person or through various media. That humans can plan, however, does not mean that all our behaviors are well thought through. Rather, human behavior is best described as a combination of planning and simpler decision-making strategies that are similar to those of other species (Kahneman 2011).

Remarkably, human mental life includes many activities that are only weakly related to behavior, such as mental arithmetic, sudoku solutions, recalling memories, and imagining possibilities. For some, mental activity is the main form of activity—for example, writers, chess players, and mathematicians. These activities are so diverse that it is difficult to summarize them effectively, yet we can highlight some common features. First, human mental processes deal extensively with sequential information, as highlighted in section 9.2.2 and especially in table 9.1. Second, human mental processes are incredibly flexible. Consider, for example, the kinds of transformations that can occur between sequences of information in the domain of language. Producing spoken language requires translation of an internal representation of meaning into a sequence of words, and understanding spoken language requires the inverse translation. Humans can master many such translations, as demonstrated by polyglots who can speak dozens of languages (Erard 2012). In addition, humans can translate to and from internal language codes through a variety of methods that are of very recent invention, such as writing systems, sign systems, Braille, and even systems of whistles (Busnel and Classe 2013). A third feature of human mental processing is that it is often time consuming (Kahneman 2011; see also section 7.7.2). We are all familiar with the time and effort it takes to make a well-reasoned decision. The extraordinary capacity of human mental processing is well exemplified by the feats of blind mathematicians who carry on lengthy computations with very little external support (Jackson 2002), by musicians who can mentally compose lengthy, multipart scores, and by chess grand masters who can play many simultaneous games blindfolded.

9.2.4 *Mental Flexibility*

Uniquely human traits probably evolved to deal with such common tasks as foraging, competition, and cooperation (see section 16.3). Once established, however, these traits gave rise to a mental revolution that is in every respect as impressive and pervasive as its behavioral and material counterparts. In addition to gathering knowledge about the physical world, humans have created elaborate mental worlds, made possible because we can freely recombine mental information without regard to whether the outcomes are meaningful in the physical world or not. This fact is perhaps most familiar in language, which can refer equally well to real and imagined events. Detailing the multitude of mental worlds created by humans would require an enormous amount of space. Here we would just like to mention a few kinds of mental worlds to give a sense of the mental flexibility that is necessary to produce them, and that appear to have no parallel in other species.

By a "mental world," we mean any set of ideas or concepts that can be reasoned about according to given rules. Mental worlds range from those that strive to adhere to physical reality, such as scientific theories, to those that make no reference to it, such as abstract mathematical constructions. Most mental worlds, however, consist of mixtures of imagined elements and elements closer to physical reality. Fascinating examples of such mixtures are found in myths, legends, and religions, but even secular belief systems can combine factual information with elements that have little connection to physical reality, such as rules of courtesy, systems of laws, and social institutions. Mental worlds tend to be internally consistent (otherwise, it would be impossible to reason about them) as well as consistent with aspects of reality—although they are typically not wholly consistent in either respect.

The construction of detailed mental worlds is a prominent aspect of fiction. Authors of fiction make no claims that their stories are true, although they typically aim at achieving a goal such as entertainment, education, or persuasion. Some fiction is created to be realistic, but often unrealistic elements are introduced, particularly in the genres of fantasy and science fiction. Authors in these genres are often praised for the care they dedicate to creating their imaginary worlds. To mention just one example, *Lord of the Rings* author J.R.R. Tolkien drew detailed maps of his imagined Middle Earth and even documented in detail the writing system, grammar, and vocabulary of several fictional languages spoken there (Shippey 2014, Fimi and Higgins 2016).

Games such as chess, Go, or sudoku also define imaginary worlds with their own rules and goals. Although some elements of games are inspired from reality (such as military jargon in chess), the rules of the game are arbitrary and have the sole goal of making the game interesting. Music and dance are also rooted in the physical reality of sound and movement, but have greatly elaborated upon these foundations by creating many distinct forms and styles—in this case turning combinatorics to their advantage, as the combinations of even small sets of musical notes and dance moves are practically endless.

In summary, the existence of myriad mental worlds, with detailed rules and rich structure, demonstrates the remarkable mental flexibility of humans as well as the centrality of sequential processing to human mental life. Games, mathematics, music, dance, stories, and so on are all examples of domains in which sequential information processing is indispensable. Although mental flexibility and sequential information processing arose in human prehistory through genetic evolution, they opened up unprecedented opportunities for cultural evolution. The result is an almost complete transformation of human mental life, which depends on culturally acquired information to a unique extent. Regardless of whether we inhabit a lush African forest, an arid Australian desert, or the frozen Arctic, we have created rich mental worlds that change and grow through cultural evolution.

9.3 Much of Human Behavior Is Evolutionarily Recent

On an evolutionary timescale, much of contemporary human activity is of very recent invention. For example, written language and agriculture are 10,000 years old or less. We do not know when humans started to count, but even mathematics deemed basic enough to be taught in middle school, such as coordinate systems and elements of set theory, is just a few hundred years old. Most advanced mathematics was invented in the last 300 years, and the same is true of modern physics, everyday technology like computers and cellular phones, and the vast majority of medical knowledge and treatments. Humans are unique among all species in that our behavior has changed extensively without intervening genetic change (Chiappe and MacDonald 2005). Since the invention of writing, detailed historical records of human culture have revealed prolonged periods of exponentially fast change (Lehman 1947, Enquist et al. 2008), with new phenomena building upon existing ones. Perhaps this is most clear for crafts and technology, but even in music and art we see clear historical trends, and expressive forms can be seen changing significantly over just a few decades. The historical development of a field often mirrors how that field is learned, such as mastering figurative art before abstract art or tonal music before atonal music (Gombrich 1977).

9.4 Uniquely Human Skills Are Learned

Detailed studies within developmental psychology and related fields have focused on the acquisition of human skills, including spoken and written language, imitation, memory, logical reasoning, cooperative skills, and motor skills in music and sports (Anderson 1982, Bandura 1986, Gibson 1999, Newell and Rosenbloom 1981, VanLehn 1996). From this wealth of information we have identified three observations that can help us understand human evolution and human uniqueness: dependence upon experience, extensive social learning, and specificity.

9.4.1 Human Skills Do Not Emerge Spontaneously

Human skills invariably require extensive training, which implies that they depend on experience rather than just on internal maturation processes. This conclusion is supported by developmental psychology (Berk 2013, Bjorklund and Causey 2017) and education science (King and Kitchener 1994, Wood 1997), as well as the study of expertise acquisition (Ericsson et al. 1993), cognitive skill acquisition (Newell and Rosenbloom 1981, Anderson 1982, VanLehn 1996), imitation (Oostenbroek et al. 2016, Heyes 2016; but see Meltzoff et al. 2018), and language development (Ellis and Large 1988, Meadows 1993, Hart and Risley 1999). Even the human skill that conceivably has the most genetic support, language, requires a remarkable amount of training (Curtiss 1977, Tomasello 2003, Bjorklund and Causey 2017). Attaining fluency in one's native language requires hearing about four million sentences (Morgan 1989) and practicing for many years. Reading and writing have a similarly slow course of acquisition (Stanovich 1986). The typical developmental trajectory of spoken and written language skills is summarized in table 9.2. It is characterized by strong cumulative effects, with each step building on previous steps. For example, a child must learn words before putting them together into sentences (Saffran and Estes 2006) and must recognize letters before she can read words (Bryant and Bradley 1985, Bryant et al. 1990). Mathematical skills also develop following a similar pattern: a child first learns numbers and the number sequence and later learns to manipulate numbers according to rules (Sarnecka and Gelman 2004, Sarnecka et al. 2007).

This stepwise, cumulative development is typical of most human skills and is the main reason why skill learning is slow and gradual: the more elements a skill consists of, the longer it takes to master (Bruner 1975, Anderson 1982, VanLehn 1996). Data also show that performance typically correlates with amount of practice, although this is not the sole determinant (Ericsson 2006). For tournament-level chess players and musicians, practice explains about 40% of the variance in performance (Ericsson et al. 1993, Charness et al. 2005, Gobet and Campitelli 2007, Hambrick et al. 2014).

9.4.2 Human Skills Are Learned Socially

The second observation is that much of the information necessary for humans to develop skills is acquired socially (Vygotsky 1962, Bruner 1966, Rogoff 1990, Tomasello 1999). Dependence upon social information is evident in the case of a skill like mathematics, which is impossible to learn on one's own. The invention of even basic features, like a positional number system and negative numbers, have required tremendous intellectual effort

Table 9.2. Examples of changes during development of spoken English and reading

Age (months or years)	Spoken language	Reading language
0–3 m	Vocal play	
6–10 m	Babbling	Discrimination between letter shapes
8–18 m	First words	
13–15 m	Expressive jargon	
13–19 m	10-word vocabulary	
14–24 m	50-word vocabulary	
13–27 m	Two- and three-word sentences	
23–26 m	Irregular and regular grammatical tenses >500-word vocabulary	
5 y	Basic skills almost fully developed	Knowledge of alphabet Verbal short-term memory Recognition of written word
6 y	Basic skills almost fully developed	Sounds associated with letters Spelling

Note: Ages are approximate, as individual development varies.
Sources: Ellis and Large (1988), Meadows (1993), Hart and Risley (1999).

over hundreds of years of painstaking conceptual development (Ifrah 1998). Similarly, language skills remain poor in the absence of an appropriate environment (Curtiss 1977), and chess learners who play in groups and receive explicit coaching perform better than players who practice alone (Gobet and Campitelli 2007). With the invention of writing and other technology for recording information, much skill acquisition is supported by artifacts like books and computer programs. The size of a chess player's library, for example, correlates with the player's skill (Gobet and Campitelli 2007). The ultimate source of this knowledge is other people, hence learning through these artifacts is also social.

9.4.3 Most Human Skills Are Domain-Specific, but Some Are General

Extensive training enables humans to become exceptionally proficient in the trained skill, but this typically has limited effect on other skills (see reviews by Anderson 1982, Ericsson et al. 1993, VanLehn 1996, Ericsson 2006, Chi 2006, Chi et al. 2014). For example, training in chess, music, or working memory does not improve mathematical and literacy skills (Sala and Gobet 2017). Extensive research about chess (Gobet and Charness 2006) highlights that skill training has multiple facets, including perception, memory, and decision making, all of which exhibit specificity of performance. For example, experts have a better memory for chess positions than novice players, not because their memory is better in general but because they have learned efficient encoding of chess positions (Chase and Simon 1973, Gobet and Simon 1996b). Further reliance on chess-specific strategies of memory encoding is apparent from studies demonstrating that chess experts remember actual game positions better than random positions that would not arise in play (Chase and Simon 1973, Gobet and Simon 1996a). When it comes to decision making, chess experts can resume play from thousands of positions they have previously studied or encountered, and are able to discard unproductive moves more quickly than beginners (Gobet and Charness 2006); but this does not necessarily improve their decision-making skills in other domains.

Transfer of performance across skills can, nevertheless, be observed between skills that share mental or motor elements (Singley and Anderson 1989, Taatgen 2013). For example, learning English is easier for Swedish than for Finnish speakers, because Swedish overlaps more with English in grammar and vocabulary (Ringbom 2007, 41). Remarkably, it is also possible to learn general skills that are helpful to solve many problems (Salomon and Perkins 1987, Perkins and Salomon 1989). These skills range from elementary information-processing operations to general problem-solving strategies. Examples of the former are list-processing operations, such as appending an item or reversing the whole list, which can be applied to any list, or counting, which can be applied to any kind of object. As examples of general problem-solving strategies, Polya's classic book, *How to Solve It*, first published in 1945, contains advice such as making sure to use all available data, looking for a related problem that is either simpler or already known, and so on (Polya 2004). Perkins and Salomon (1989) review evidence that these and similar strategies help experts in many fields solve unfamiliar problems. They also provide a detailed analysis of transfer of expertise and examples of both successful and unsuccessful transfer. We consider how human mental skills are learned in chapter 13.

9.5 Explaining Human Uniqueness

A successful theory of human evolution must explain how humans have gained extensive behavioral and mental flexibility, given that the behavioral and mental skills of other species are much more constrained genetically. In turn, this requires explaining how humans have been able to circumvent

Table 9.3. Characteristics of hypotheses about human evolution

	Solutions to new problems	Extensive behavioral change on a historic timescale	Extensive cultural evolution of mental skills	Skill learning is slow and gradual	Skills may generalize across domains	Skill learning depends on social information
Evolutionary psychology	No	No	No	?	No	?
Inborn general intelligence	Yes	?	No	?	Yes	?
Cultural evolution	Yes	Yes	?	Yes	Yes	Yes

Note: A question mark indicates that the hypothesis is not spelled out in sufficient detail to derive a clear prediction.

the exponentially high learning costs of behavioral and mental flexibility (chapters 4, 8, and 11).

Perhaps the most difficult question is, Why only humans? What has prevented the evolution of humanlike abilities in other species? We consider this question in chapter 16. In this section, we summarize the pros and cons of some major theories of human evolution, with a focus on the three characteristics of human behavioral and mental abilities reviewed in section 9.4. Table 9.3 summarizes some characteristics of these hypotheses. We can confidently reject and exclude from further discussion two extreme hypotheses. First, human uniqueness cannot be entirely a product of genetic evolution, because much of what makes humans unique requires extensive social input and is of recent invention (section 9.2). Second, human uniqueness cannot be entirely a product of cultural evolution, because, for example, apes raised in human environments do not become cognitively human (Kellogg and Kellogg 1933, Terrace et al. 1979). A key task of a theory of human evolution is to determine the roles of genetic and cultural evolution, as well as of physical and ecological factors.

9.5.1 *Evolutionary Psychology*

The core idea behind evolutionary psychology is that the human mind has evolved to contain many specialized mental modules, each of which is hardwired to solve specific problems (Tooby and Cosmides 1989, 1992, Cosmides and Tooby 1992, Pinker 1994, Buss 2005, Rozin 1976; for critique, see Buller 2006). Human behaviors that have been attributed to such specific modules include child care (Daly and Wilson 1998), jealousy (Daly and

Wilson 1988), mate choice (Buss 2000), and social intelligence or theory of mind (Whiten and Byrne 1988, Whiten 1997, Cosmides and Tooby 1994, Cosmides et al. 2010). Similar views are shared by other nativist theories of human faculties. Some linguists, for example, think of grammatical abilities as an "instinct" similar to other complex genetically programmed traits, such as web weaving in spiders and dam building in beavers (Chomsky 1980, Pinker 1994). Likewise, some neuroscientists have argued for an innate imitation ability (Rizzolatti and Craighero 2004, Iacoboni and Dapretto 2006).

According to evolutionary psychology, human mental modules have evolved through natural selection in order to deal with problems that were common in our evolutionary past (Cosmides et al. 1992, Cosmides and Tooby 1994). The modular organization of the mind would be common to all animals, but humans would possess a much larger number of mental modules (Cosmides and Tooby 1994, Buller 2006, Burkart et al. 2017). Thus many evolutionary psychologists reject the idea that culture can shape human cognition, and hold instead that culture is simply a manifestation of the operation of innate mental modules (Tooby and Cosmides 1989).

The view of the mind as a collection of inborn mental modules has appeal. For example, it provides a conceptual framework for how genetic evolution influences human behavior (by creating mental modules), and it identifies both what unites animal and human minds (the modular architecture) and what separates them (the number of modules). Indeed, much research in evolutionary psychology has dealt with classic topics of evolutionary biology rather than with uniquely human phenomena. An additional rationale for a modular mind is of prime importance to our book. Both evolutionary psychologists and nativist linguists recognize that combinatorial dilemmas are a major obstacle to learning and development. For example, Tooby and Cosmides (1992, 101) write: "Designs that produce 'plasticity' can be retained by selection only if they have features that guide behavior into the infinitesimally small regions of relatively successful performance with sufficient frequency." This argument also implicitly recognizes that different problems have different solutions; otherwise, just a few mental modules would suffice. Similarly, the "poverty of the stimulus" argument in linguistics recognizes the problem of learning a language from environmental input (Chomsky 1980). A modular mind seems a straightforward solution to these problems: if important mental skills are innate, learning just needs to fill in the details (Fodor 1983, Tooby and Cosmides 1992, Gelman and Williams 1998, Chiappe and MacDonald 2005).

Despite its appealing features, evolutionary psychology is hard to reconcile with the fact that characteristically human traits require extensive and specific training in order to develop (section 9.4), and with the fact that many of these traits are too recent to have a specific genetic basis (section 9.3). For example, how could we have innate modules for driving cars, programming computers, composing a symphony, or playing chess? While it is possible to consider a module for domain-general learning and planning, or a module for combining the abilities of other modules (discussed next), such modules would end up being responsible for the vast majority of uniquely human behaviors, which would undermine the premise that human uniqueness derives from a host of specialized modules. It seems to us that the facts summarized in sections 9.2–9.4 point to a much larger role of acquired information than can fit under the umbrella of evolutionary psychology.

It has also been suggested that the foundation of human cognition lies in *a few* inborn modules, called *core knowledge systems*. For example, Spelke and Kinzler (2007) consider five modules: objects, actions, numbers, space, and social partners. Similarly, Carey (2009) considers core systems for agents, numbers, and objects, the latter including also causal and spatial relations. Core knowledge systems are assumed to be inborn (similar to other notions of cognitive modules), but also to serve as the foundation for learning further mental skills. This view is similar to ours, although we focus on even more basic abilities related to sequential information processing and we consider in more detail how culture can contribute to mental skills.

9.5.2 Inborn General Intelligence

That humans have an inborn ability to tackle practically any problem has been suggested in many disciplines, such as archaeology, anthropology, behavioral ecology, neuroscience, and philosophy (Fodor 1983, Donald 1991, Klein 1995, Mithen 1996, Klein 2002, Chiappe and MacDonald 2005, Gärdenfors 2006, Sol 2009, Barrett 2015). Some of these theories assume a partly modular cognitive architecture, but, in contrast with evolutionary psychology, they focus on general intelligence rather than on domain-specific adaptations (Guilford 1967, Fodor 1983, Sternberg et al. 2002, Baum 2004, Gardner 2011). Many authors, for example, have proposed that humans solve problems through mental simulations of possible scenarios, in order to predict future events and compute a course of action (Craik 1943, Tulving 2002, Hesslow 2002, Suddendorf and Corballis 2007).

The hypothesis that humans have an advanced mechanism for general problem solving provides an intuitive account of human uniqueness that is compatible with the flexibility and proficiency of human cognition (section 9.2). This hypothesis, however, leaves two important questions open. The first is the origin of the information upon which the general problem-solving mechanism would operate. For example, even the best route-finding algorithm cannot bring you to your destination without knowledge of the specific transportation infrastructure in your area. As it is unrealistic to assume that each individual independently develops all the knowledge underlying uniquely human skills, a general intelligence hypothesis seems viable only if complemented with extensive amounts of individual and social learning (see chapter 11). Reliance on learning could explain both gradual skill development and specificity of performance, and would therefore be compatible with the basic facts of human intelligence.

The second open question is whether the general problem-solving mechanism can evolve by natural selection. Sometimes intelligence is treated as a one-dimensional quantity, such as body mass or height, that natural selection can increase or decrease (Deary et al. 2010, Nisbett et al. 2012), but intelligence depends upon a variety of abilities, like working memory, processing speed, logical reasoning, and imagery (Mackintosh 2011, Ritchie 2015, Ramus 2017). Without a detailed model of how such abilities operate together, it is difficult to study the evolution of intelligence. Some insight, however, comes from the field of artificial intelligence (AI), which historically had placed general intelligence among its main goals (Nilsson 2009, Ohlsson 2012). For example, the General Problem Solver created by Newell et al. (1959) attempted to solve problems by looking for a sequence of operations that would transform an initial state into a final state with desired properties. In this system, proving a mathematical theorem would begin with a set of axioms (initial state) and admissible rules of inference (operations), with the goal of finding a sequence of inferences arriving at the theorem (final state). Because complex problems require long sequences of operations, it was soon realized that general problem solving encounters the same combinatorial difficulties as animals attempting to learn long behavioral sequences. That is, solving times increase exponentially with problem complexity. AI was thus redirected toward finding efficient solutions in restricted problem domains, using "expert systems" endowed with as much domain-specific knowledge as possible (Newell et al. 1972, Feigenbaum 1977, Hayes-Roth 1977). For example, the general theorem-proving program might be replaced by a Euclidean geometry program that could access a library of already proven statements and employ known theorem-proving strategies. Such a program would fare better than the General Problem Solver in the domain of Euclidean geometry, but would be unable to work outside this domain. We come back to this argument in chapter 11.

Although AI has progressed tremendously from its early days, the basic insight still holds that even the most efficient algorithms are powerless without domain-specific information. Indeed, this was the main reason why evolutionary psychologists deemed the evolution of general intelligence unlikely, and instead sought to partition intelligence into domain-specific modules (section 9.5.1). Our argument about animal intelligence has a similar starting point—combinatorial difficulties—but a somewhat different conclusion. We argue that the main mechanism of animal intelligence (i.e., the chaining algorithm) is domain-general, but also that combinatorial difficulties are circumvented by focusing on relatively simple problems (short sequences) and by encoding domain-specific knowledge in genetic predispositions (chapter 4). Regarding human intelligence, it seems to us that neither evolutionary psychology nor inborn general intelligence are wholly satisfactory hypotheses.

9.5.3 *Cultural Evolution*

That human uniqueness arises from cultural processes is the classic view within the social sciences and humanities. Many theories of human culture and how it changes have been formulated and have yielded important insights (e.g., Kuhn 1996, Munro 1963, Berger and Luckman 1967, Steward 1972, Geertz 1973, Harris 1999, Carneiro 2003). In the 1970s, the traditional conceptual approach to culture was complemented by a research program emerging from evolutionary biology, which sought to develop a mathematical theory of cultural evolution (Cavalli Sforza and Feldman 1981, Lumsden and Wilson 1981, Boyd and Richerson 1985, Laland and Brown 2011). These efforts have produced a rich body of theory describing how low-level processes, such as transmission, retention, and innovation, can create cultural phenomena over time (Acerbi et al. 2014, Henrich 2015, Mesoudi 2016, Creanza et al. 2017).

Focusing on culture as the determinant of human uniqueness easily fits the basic facts highlighted above: that many human skills have a recent origin, that they require extensive training, that they are learned socially, and that they are very diverse both over time and in different societies. It is more challenging to articulate a comprehensive theory of how cultural evolution interacts with genetic evolution to create human intelligence. We cannot explain human uniqueness just with cultural evolution, as there must be some inborn mechanism—absent in other species—that allowed humans to become cultural beings. Understanding these inborn factors is of paramount importance to understanding human uniqueness, the scope and power of cultural evolution, and the origin of human culture. Indeed, theories of cultural evolution often explain human uniqueness by assuming that crucial elements of human psychology are innate. For example, Boyd and Richerson (1985) proposed that inborn psychological biases, such as preferential learning from successful individuals or from the majority, are crucial for culture to accumulate productive behaviors, despite the fact that unproductive behaviors could be transmitted equally easily (see also Richerson and Boyd 2005, Eriksson et al. 2007, Henrich 2015).

Recently, theorists have suggested that cultural evolution may have a broader reach than has often been assumed, allowing humans to learn and transmit not only behaviors and material culture but also mental abilities (Gabora 1997, 1998, Richerson and Boyd 2005, Tomasello 2014, Henrich 2015, Heyes 2018). This broadening of the role of cultural evolution offers new and powerful explanations of human uniqueness, and brings theorizing from biology and psychology closer to traditional views in the social sciences and the humanities. We briefly survey some of these ideas.

Tomasello and coworkers have suggested that thinking already exists in primates, where it is mainly used in social competition. However, they posit that humans have unique inborn abilities and motivations to engage in collaborative endeavors, such as attending jointly to the same stimuli and behaviors, understanding others' intentions, and formulating collective goals (Tomasello 1999, Tomasello et al. 2005, Herrmann et al. 2007, Tomasello 2014; see also Frith and Frith 2010, 2012, MacLean 2016). In this view, these factors are the basis of human cooperation and cultural transmission, including transmission of mental abilities, such as language. In this view, the role of cumulative culture is so important that, without an appropriate social environment, humans would be very similar to apes (Tomasello and Rakoczy 2003).

Gabora (1997, 1998) has proposed a concrete mechanism for the transition from a world poor in culture to one in which culture is a major force, which is similar to the ideas we present in chapter 16. In this scenario, the human mind is very open but initially devoid of complex mental content. Innovation proceeds by creating and recombining cultural elements, but this process is slow until the cultural system reaches a critical threshold in complexity and size. Past this threshold, the generation of new mental content increases dramatically due to the availability of both raw mental material and mental mechanisms that can recombine this material into new forms. Gabora and Steel (2017, 2020) have studied this process using mathematical models developed in theoretical biology to study the emergence of complex chemical networks early in the evolution of life.

Finally, Heyes's (2018) view of the role of culture in human cognition is very similar to ours. Based on an eye-opening review of empirical data, she concludes that social interactions are crucial for the development of abilities that traditionally have been considered largely innate, such as grammatical competence, imitation, and theory of mind. Further, she suggests that these abilities are themselves largely the product of cultural evolution, and that they can be learned effectively without strong genetic support based on domain-general learning processes of sequential learning and associative learning. In the following, we reach a similar conclusion based on theoretical arguments and a complementary review of the literature.

10

The Transition

- The human evolutionary transition set humans apart from other species and resulted in cultural evolution on a grand scale.
- The transition was made possible by genetic changes enabling humans to faithfully represent and process sequential information, and to learn mental skills beyond what our genes provide.
- These genetic changes did not immediately make humans different from other animals, but they resulted first in gene-culture coevolution and then in extensive cultural evolution.
- Most of the differences between humans and animals arise from cultural evolution, resulting in a diversity of uniquely human mental and behavioral skills, such as thinking and extensive social learning abilities.
- Natural selection typically works against mental and behavioral flexibility because of the associated combinatorial costs. In humans, these costs are offset by relying on culture as a reservoir of mental and behavioral skills, enabling individuals to learn skills that lie far beyond what they could learn on their own.

10.1 Introduction

Throughout the book, we have emphasized that discovering productive behavioral sequences is hampered by severe combinatorial dilemmas, such that complex behavioral skills cannot be learned within a realistic time frame without prior knowledge or facilitating circumstances. In chapters 3–8, we argued that animals tackle this problem mainly by using genetic predispositions to guide learning, and by learning only single responses or short sequences, often aided by favorable entry patterns provided by the ecological or social environment. This strategy reduces combinatorial problems and enables animals to learn functional behavior, but it also restricts what can be learned. In this and the following chapters, we argue that, during human evolution, a new, eventually more effective strategy has emerged to tackle combinatorial dilemmas: to rely on cultural rather than genetic knowledge. This evolutionary transition fundamentally changed human evolution (chapter 1; see also Maynard Smith and Szathmáry 1995).

Our hypotheses regard the evolutionary events that made this new strategy possible. In brief, we make two claims about the human mental architecture. The first is that there are very marked mental differences between humans and other animals. We believe this claim to be well supported by the evidence reviewed in chapters 3–8, indicating that animal intelligence is satisfactorily described by a synthesis of classical ethological and psychological concepts; in other words, as associative learning guided by genetic information (hypothesis 2 in chapter 2). Human skills, on the other hand, lie far beyond what can be thus achieved (chapter 9). Our second claim is that what sets humans apart from other animals is overwhelmingly learned rather than genetically determined. For example, we suggest that our abilities to plan, imitate, and cooperate would not emerge without extensive learning within a social environment (Heyes 2018). Further, we propose that social input not only contributes to the development of behavioral skills, but it also results in the acquisition of mental skills that have evolved culturally (hypotheses 4–6 in chapter 2). A complex feedback exists between cultural evolution and human cognition, which we attempt to unravel in this and the following chapters.

10.2 How Human Evolution Is Different

We suggest that, during human prehistory and after the split between human and chimpanzee lineages, a few decisive genetic changes occurred that resulted in novel general-purpose abilities, namely, to represent sequential information and to learn new ways to process information. These genetic changes, however, did not immediately make us different from other animals. Rather, mental and behavioral divergence from animals occurred via the ensuing cultural evolution.

It is crucial to our argument to distinguish between uniquely human mental and behavioral skills, which we argue are predominantly learned, and the genetic foundation that permits such learning. Using a computer metaphor, we can compare genetic foundations to the processor and its memory, and behavioral and mental skills to software that can run on the computer. We envision that, when the human brain assumed its current genetic makeup, its "software" was not very different from that of other species. The potential of the new processing power remained largely unexpressed until cultural evolution devised better and better software for mental and behavioral skills. For this to be possible, it is necessary that mental skills be socially transmitted, as we detail below and in subsequent chapters. It is also necessary that the initial genetic changes conferred some direct individual advantage, so as to be preserved by natural selection. For example, a better memory for sequences of events may have improved even a basic associative learning system by providing a more accurate source of information (chapter 5). We return to the important question of these initial changes below and in chapter 16. Here we want to stress that our hypothesis about genetic changes (hypothesis 5, chapter 2) restricts these changes to a few innovations in general-purpose information processing. In summary, we propose the following:

- *Everything uniquely human derives from a few genetically based changes, including (1) changes in memory functions, yielding the ability to store and process sequential information faithfully, and (2) changes in mental flexibility, yielding the ability to learn new mental skills. These abilities are absent, or rudimentary, in other animals.*
- *These genetic changes sparked cultural evolution, eventually resulting in additional, uniquely human mental skills that now pervade our behavioral and mental machinery, in addition to preexisting skills, such as associative learning.*

This picture of human cognitive evolution is motivated, first of all, by the fact that there must exist genetic differences in mental processes between humans and other animals; otherwise, other species could learn everything we can learn. Our main reason for positing that these changes were few is parsimony, combined with the realization that cultural evolution can be the main force driving mental evolution once it becomes possible to learn new mental skills.

10.3 Genetic Foundations of Human Uniqueness

What were the genetic changes that eventually allowed humans to think, speak, and create culture on a grand scale? We believe that a strong candidate is a change in memory that made possible the faithful *representation and processing of sequential information*. We refer to this ability as *sequential memory*, but it would also be possible to describe it in terms of the traditional psychological concepts of working, semantic, and episodic memory. As argued in sections 5.4.2 and 9.2.2, a faithful sequential memory is necessary to support human abilities and appears to be lacking in other species. To enable human mental abilities, sequential memory must apply to both current information (working or short-term memory) and to the use of stored information for planning, from minutes to years ahead (long-term and episodic memories). Likewise, it must apply to both external and internal events, such as when we recall the steps of a previously formulated plan. While we are confident that enhancements to sequential memory have occurred, we cannot pinpoint exactly what they have been. For example, the genes may provide us with a memory with high temporal resolution, but learning may be necessary to fully use its potential. That one-year-old children can replicate sequences of two or three actions suggests strong genetic support for sequential memory (Bjorklund and Causey 2017), although we also know that memory for events and for sequences improves during childhood through social learning facilitated by parents and caregivers (Nelson and Gruendel 1981, Nelson and Fivush 2004, Bjorklund and Causey 2017). Childhood development of sequential abilities is considered in chapter 13.

The second feature that we believe to be necessary for human abilities is *flexibility of mental processing*. The changes that make this ability possible are even harder to pinpoint than for sequential memory, yet the ability can be described fairly simply in terms of information processing. By mental flexibility, we mean the ability to learn mental procedures that operate on mentally represented information. We can describe as follows a series of mental procedures that intervenes between a stimulus S and a behavior B:

$$S \to P' \to M' \to P'' \to M'' \to \ldots \to M^{(n)} \to P_{\text{behavior}} \to B \qquad (10.1)$$

where the Ps are mental procedures, the Ms are intermediate memory states that represent the results of mental processes, and P_{behavior} is a final process that translates the current memory state into behavior. (It is also possible to consider multiple parallel streams of mental procedures.) For example, you can imagine going through a series of mental steps when calculating $128 + 128$. The description in equation 10.1 follows the classical "cognitive architecture" approach to human cognition, which seeks to understand how information is transformed by successive mental operations (VanLehn 1996, Ericsson 2006, Chi et al. 2014, Anderson 1993). The point we wish to make here regards the *origin* of the Ps. In a behavior mechanism that is genetically determined, the Ps and their succession are essentially fixed. For example, the chaining model (chapter 3) can be described with just two mental procedures that operate on short sequences of the form $S \to B \to S'$. One procedure updates memory (v and w values) based on its current state M, and the events S, S', and B:

$$(M, S, B, S') \to P_{\text{memory}} \to M'$$

where M' is the resulting memory state (the updated v and w values). The other procedure uses the current stimulus and memory state to generate behavior:

$$(S, M) \to P_{\text{behavior}} \to B$$

A complete model of animal intelligence would include a few more procedures, for example, for stimulus recognition and motivation. The main idea, however, is that all procedures develop from a strict genetic blueprint. Experience may be necessary for proper development, such as in sensory processing, but it cannot create new ways of processing information. In a mechanism with mental flexibility, on the other hand, the *P*s and their potential combinations are learned and subject to change. Such learning may include how stimuli are represented, including symbolic codes, how to update memory, and how memories are used to generate behavior. Chapter 13 is dedicated to this topic.

Motor flexibility, which is also highly developed in humans, can be viewed as a particular instance of mental flexibility. In most animals, motor patterns are genetically programmed (chapter 4), but humans and a few other species exhibit a high degree of motor flexibility. Mental and motor flexibility have similar costs and benefits. Just as a larger behavioral repertoire may allow for better exploitation of the world, an increased repertoire of mental procedures can provide a better toolbox for solving problems (Karmiloff Smith 1992, Ericsson et al. 2018). In both cases, however, a larger repertoire introduces substantial learning costs (chapters 4 and 15). The interplay between motor flexibility and cultural evolution is examined in chapters 15 and 16.

We also speculate that genetic changes have modified the human *motivation for learning and social interaction* (Tomasello et al. 2005). Whereas most animals appear to give up relatively rapidly when they explore a new problem and are met with failure, humans tend to persevere. Language learning is a good example: It takes many years, countless hours of practice, and the correction of many mistakes (Morgan 1989). Far from being an effortless task, it requires a remarkable amount of motivation to master such a difficult skill. Within our framework, it is expected that a species that focuses on learning of long sequences should display perseverance in learning, because long sequences are challenging to learn. Likewise, a strong motivation for social interaction is expected if other individuals are an important source of knowledge.

The literature contains many proposals of genetic differences in information processing and memory that may distinguish humans from other animals, but there is little agreement as to which differences are crucial. Some authors hypothesize that human thinking derives from enhanced working memory (Wynn and Coolidge 2004, Coolidge and Wynn 2005). Our hypothesis about genetic changes is similar in that both sequence processing and mental flexibility require an efficient working memory. Other proposals for distinct genetic differences include unique planning abilities (Suddendorf and Corballis 2007), the capacity for symbolic thought and language (Fodor

1975, Deacon 1997), recursive thought (Hauser et al. 2002, Corballis 2011), and theory of mind (Tomasello 1999). These proposals are similar to ours in that they suggest that one or a few specific changes may have large consequences. However, we emphasize that the genetic changes that make humans unique may have been domain-general and almost behaviorally silent initially, realizing their full potential only through cultural evolution. Gabora (1997, 1998), Tomasello et al. (2005), and Heyes (2018) have suggested a similar scenario (section 9.5.3).

10.4 Cultural Evolution of Human Uniqueness

The idea that many human phenomena are the product of cultural evolution is a mainstay of the social sciences and the humanities (Carneiro 2003). Our hypothesis about cultural evolution (hypothesis 6, chapter 2) goes perhaps further than most in maintaining that culture not only provides content for human mental life: it also shapes the workings of the human mind, to the extent that even fundamental abilities, such as thinking, planning, social learning, and cooperation, may not develop without cultural input. The work of Vygotsky (1978) contains influential early explorations of the idea that human mental processes owe much to social learning. More recent discussion of the role of culture in human cognition can be found in Jablonka and Lamb (2005), Tomasello (1999), Prinz (2012), and Heyes (2018). To this work, our hypothesis adds a theoretical justification for why human cognition is predominantly cultural. As stated in the introduction to this chapter, our idea is that the combination of culture and extensive mental flexibility is an alternative, and more powerful, strategy to address combinatorial dilemmas, compared to the combination of individual experiences and genetically guided associative learning that characterizes other animals. In other words, humans could not be mentally flexible if they did not have culture, and culture could not exist on the same scale if human mental procedures were hardwired. This aspect of our hypothesis is discussed in chapters 15 and 16, and is inspired by the tradition of mathematical modeling of cultural evolution started by Cavalli Sforza and Feldman (1981) and Boyd and Richerson (1985).

10.5 Gene-Culture Interactions

Genetic and cultural evolution are not completely separate in time, and it is likely that simple culture and genetic abilities for culture have reinforced each other during an extensive period of coevolution (Simoons 1970, 1971, Durham 1991, Cavalli Sforza et al. 1996, Richerson and Boyd 2005,

McElreath and Henrich 2007, Richerson et al. 2010). According to our hypotheses, human cognitive evolution became largely cultural once the ability to learn new mental procedures was in place, but this does not exclude enduring genetic influences. For example, cultural evolution is more likely to establish mental and behavioral traits that fulfill genetically programmed motivational and emotional needs. At the same time, cultural evolution can modify even supposedly hardwired goals. For example, while we expect reproduction to be highly constrained genetically, cultural evolution has produced many uniquely human phenomena in this domain, such as religious abstinence, honor killings, and arranged marriages.

Gene-culture interactions are also evident in observations of heritable genetic variation in culturally evolved mental procedures, such as chess playing or musical abilities (Plomin and Petrill 1997, Deary et al. 2006). These observations reveal that the genes influence human abilities to learn and use culturally evolved mental procedures. For example, working memory and motivation to learn exhibit heritable genetic variation (Ando et al. 2001, Luciano et al. 2001, Polderman et al. 2006, Kovas et al. 2015). The observations, however, do not imply that the genes influence mental abilities directly. For example, two students who are identical in all respects but their genetic makeup may perform differently in calculus class, but this does not prove that the mental procedures that underlie calculus are hardwired. Furthermore, genetic variation might influence mental procedures in one cultural environment but not in another. For example, a genetic variant may be associated with worse reading proficiency when schooling is poor, but not when reading instruction is good.

10.6 How Our Hypotheses Explain Human Uniqueness

Our hypotheses attribute a major role to social experiences in determining not only the content of the human mind but also its workings. Further, they contend that culture acts as a repertoire of mental abilities, in addition to its accepted role as a repertoire of knowledge. Does this picture agree with the observations about human behavioral and mental skills summarized in chapter 9, and specifically those in table 9.3? We can easily account for the fact that *much human behavior is evolutionarily recent,* since we are assuming that human behavioral and mental skills are primarily the product of cultural evolution. This is also consistent with the fact that *uniquely human skills take time and effort to learn.* Consider a skill that requires mastering a mental sequence in which the initial information is transformed by a number of *P*s, for example, arithmetic addition. According to our hypotheses, the necessary *P*s would not be available early in development, such as the *P*s to transform visual or auditory input into mental representations of numbers and the *Ps* to manipulate addends to get their sum. If the skill is complex, its constituent *Ps* may themselves depend on other *Ps* that must also be learned. This implies that development has a typical course corresponding to the order in which *Ps* are learned, ranging from simple to complex. However, unrelated skills relying on different *Ps* can develop in parallel, with no particular temporal relation to one another. Depending on the effect of each *P* on overt skill, developmental trajectories could display either gradual change or distinct stages. The former would occur when each *P* has approximately equal impact on skill, the latter when learning certain *Ps* results in big jumps in skill. Similar reasoning applies to the acquisition of motor skills.

The developmental process we have just outlined is compatible with theories from developmental psychology (Case 1978, Karmiloff Smith 1992, Thelen and Smith 2006, Munakata 2006, Fischer and Bidell 2007, Bjorklund and Causey 2017). It can also accommodate variation in individual developmental trajectories, to the extent that a flexible order is possible when acquiring *Ps*. Our hypotheses also account for the *strong dependence of development on social information* (Nelson and Fivush 2004, Bjorklund and Causey 2017), because a complex set of *Ps* cannot be invented by an individual on its own. This topic is discussed in chapters 12 and 13. Finally, our hypotheses predict *specificity of performance* in most skills. Such specificity would arise to the extent that different skills require different *Ps* or combinations of *Ps* that need to be learned separately. In the following chapters, we consider in more detail how *Ps* can be transmitted between individuals and how cultural evolution can leverage this ability to build effective mental and behavioral repertoires.

11

How and Why Does Thinking Work?

- Thinking requires representing sequential information and learning new mental skills.
- We study a model of thinking based on a domain-general ability to learn mental models as well as domain-specific mental skills.
- In ecological niches typical of nonhuman animals, thinking may not confer an advantage over associative learning.
- However, thinking outperforms associative learning when combined with cultural information.
- Cultural information can include knowledge of the world as well as domain-general and domain-specific mental skills.

11.1 Introduction

In this chapter and the three that follow, we explore some important consequences of the genetic abilities that we suggest started the human evolutionary transition: faithful sequence representation and mental flexibility. These abilities have profound consequences for the behavioral and mental skills of humans and for the cultural evolutionary process. This chapter shows how sequential processing abilities can be used in thinking, and how thinking requires mental flexibility. At the same time, we stress that these new abilities are not superior to associative learning in all environments. Considering costs as well as benefits can offer important clues to the evolution of thinking, because it narrows down the environments in which such evolution is likely.

11.2 Thinking as Mental Sequencing

In section 2.5, we defined thinking as the ability to mentally recombine information acquired from different experiences to discover new, productive sequences of actions. We call this ability *mental sequencing*. This definition of thinking is broad enough to encompass a diversity of mental mechanisms. Here we focus on mechanisms that build a mental model of the world and explore it mentally. These mechanisms are central to the debate around animal intelligence (chapter 7) and also capture how the terms *planning, reasoning*, and *thinking* are used in cognitive psychology (Johnson-Laird 2006, Gärdenfors 2006, Kahneman 2011). From this perspective, a satisfactory understanding of thinking requires understanding both how the mental model is built and how it is used. Discussions of thinking often focus on the latter (how deductions are made or how problems are solved), yet mental model building is equally crucial. In this section, we introduce the basic ideas behind mental models. In the following sections, we present a mathematical model of thinking and use it to explore the costs and benefits of thinking, especially in relation to chaining.

11.2.1 The Mental Model

The idea behind mental models is to learn continuously about causal relationships in the environment, even if they do not appear immediately meaningful, in the hope that this information becomes useful later (see also section 7.4). A simple way for a learner to construct a mental model is to store experienced stimulus-behavior-stimulus triplets in memory. These sequences are written as $S \to B \to S'$ in our notation and are also known as stimulus-response-outcome associations in experimental psychology (Pearce 2008, Bouton 2016). For example, the experiences

1. *Full bottle* → Pour → *Liquid*
2. *Liquid* → Drink → *Sweet*

can be stored in memory and retrieved later to decide on how obtain the reward *Sweet*.

Learning a mental model is demanding. First, it requires accurate perception and representation of $S \to B \to S'$ sequences, and possibly of longer sequences if longer-term relationships are to be encoded. Second, it requires a large memory; much larger than, say, just storing stimulus-response values or associations. Finally, it requires mental mechanisms to update the model based on experience, which can be complicated by the presence of uninformative stimuli and by the same action in the same situation potentially having different outcomes. For example, striking a nut with a stone can crack the nut open, but it can also have no effect or cause the nut to fly away.

11.2.2 Mental Sequencing

The goal of learning a mental model is to use it later for mental sequencing, that is, to plan sequences of actions that can take the individual from its current situation to a goal (Miller et al. 1960, Anderson 1993, Johnson-Laird 2006). A simple way to do mental sequencing is to search memory for $S \rightarrow B \rightarrow S'$ triplets whose last and first stimuli match. For example, this would allow for combining information from the triplets above to infer that pouring from the bottle is conducive to accessing its sweet contents. When the memory contains many triplets, longer action sequences can be planned, but doing so becomes more demanding, as we discuss below.

11.2.3 Exploration

Mental sequencing can find the best plan according to the current mental model, but exploring the environment instead of following this plan could reveal even better options. As discussed in chapters 1 and 3, any decision-making mechanism must strike a balance between using current knowledge and looking for additional knowledge (sections 1.3.1 and 3.2). Decisions based on thinking are no exception. Indeed, exploration is even more important in thinking, because the potential advantages of thinking rely on having learned a mental model that is sophisticated enough to reveal action sequences that cannot be discovered by simpler mechanisms, such as chaining.

11.3 When Does Thinking Pay?

There are at least three areas in which mental sequencing can be more efficient than chaining. First, sequences of actions can be learned faster by combining different experiences, regardless of their order, whereas chaining must wait for the experiences to occur in the correct sequence (figure 11.1A). Second, thinking can adapt more quickly to changes in the external world. If a plan is foiled, such as when a known route becomes inaccessible, a new route can be planned using the mental model, rather than learned by trial and error as in chaining (figure 11.1B). Similarly, a mental model can be learned in the absence of rewards and can be used to locate rewards quickly when they are introduced (section 7.4). Third, learned mental models can be used to solve new problems. For example, a mental model of a neighborhood can be used to deliver packages to a different set of houses each day (figure 11.1C).

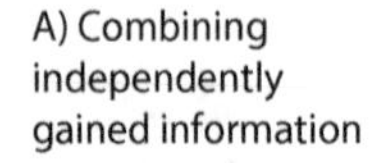

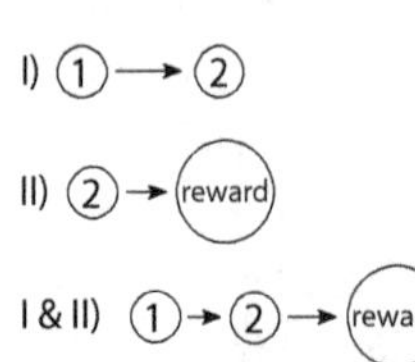

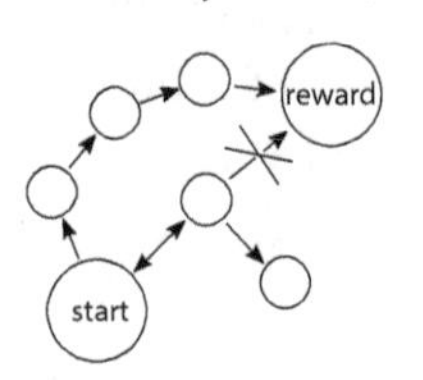

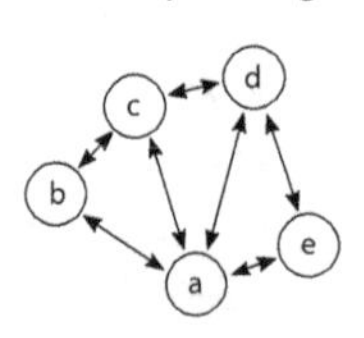

FIGURE 11.1. Examples of problems that can be solved with mental models but not with chaining. In all cases, the learning of new information requires modification of a previously established response. In contrast to chaining, decision making based on a mental model can adapt immediately to these new circumstances. A: The model learns how to reach state 2 from state 1, then learns that a reward can be obtained via state 2. B: The shortest path to the reward is suddenly blocked. C: A new path is required each time to fulfill novel goals in a familiar world, such as traveling from a to b passing node c.

These general advantages notwithstanding, to understand the evolution of thinking we need to understand in detail when it performs better than chaining. For example, thinking offers no benefits when it comes to learning a single behavior (sequences of length one), because in this case there is no opportunity for mental sequencing. Thinking can perform better than chaining when learning sequences of two or more actions, but understanding the magnitude of this advantage requires quantitative modeling. In the next section, we introduce a quantitative model of thinking and compare it to chaining in several problems. We hope to show that thinking can carry large costs, both in terms of the required mental machinery and in terms of learning time. These costs imply that thinking is not invariably superior to associative learning. However, we also argue that thinking can greatly outperform chaining when mental models can be learned socially and evolve through cultural evolution. Indeed, thinking can be even more effective if mental skills for building and using mental models can themselves evolve culturally. The power of thinking and its magnification through cultural evolution is explored further in the upcoming chapters.

11.4 A Mathematical Model of Thinking

In this section, we introduce a simple mathematical model of thinking that operates according to the principles put forward in the previous section. The goal of this model is twofold. First, we want to show that a mental mechanism capable of thinking must necessarily be quite a bit more complex than an associative learning mechanism. As we detail in section 11.6, this complexity comes with costs. Second, we want to study how powerful thinking is compared to chaining. To fulfill these goals, we formulate a model that is conceptually simple and at the same time capable of finding optimal sequences of actions in the problems we consider. We emphasize that the model is not meant to be fully realistic, but rather informative about the power of thinking. We see in sections 11.6 and 11.7 that a real-world thinking mechanism must depart in many ways from our model.

11.4.1 Learning a Mental Model

The first step in a model of thinking is spelling out how mental models are learned. Our model starts from experienced $S \to B \to S'$ triplets. Such an experience provides evidence that using B in response to S leads to S' and also that it does *not* lead to other stimuli. In general, the probability $z(S \to B \to X)$ that X follows $S \to B$ can be estimated using the following learning rule:

$$\Delta z(S \to B \to X) = \alpha_z \left[\lambda_X - z(S \to B \to X')\right] \tag{11.1}$$

where α_z is a learning rate and λ_X is the target value of $z(S \to B \to X)$. That is, $\lambda_X = 1$ for the stimulus that actually occurred ($X = S'$), and $\lambda_X = 0$ for all other stimuli ($X \neq S'$). Equation 11.1 is very similar to the learning equations that our chaining model uses to learn stimulus-response values and stimulus values (equations 3.11 and 3.12). Over time, it brings $z(S \to B \to X)$ close to the average experienced value of λ_X. Because λ_X is 0 when X does not occur, and 1 when it does, the average of λ_X is the probability that X occurs after $S \to B$, which is, in fact, what we are trying to estimate. For example, if X always follows $S \to B$, we have $\lambda_X = 1$ always, and $z(S \to B \to X)$ will approach 1. On the other hand, if X follows $S \to B$ only half the time, λ_X will be 0 and 1 equally often, and $z(S \to B \to X)$ will approach $1/2$. In conclusion, by having a z variable for each experienced $S \to B \to S'$ triplet, we can learn a mental model that embodies causal information about what outcomes follow each behavior in any given situation. Equation 11.1 can be extended to stimuli composed of many elements, in a similar way as customarily done in associative learning models (section 6.4; see also Enquist et al. 2016, Ghirlanda et al. 2020). For simplicity, we do not consider this extension.

11.4.2 Mental Sequencing

The mental model can be used for mental sequencing by looking for sequences of actions that link the current stimulus with other stimuli. For example, if both $z(S \to B \to S')$ and $z(S' \to B' \to S'')$ are greater than 0, it should be possible to go from S to S'' by first using B and then B'. Furthermore, according to the mental model, the probability that S'' is actually reached is the product $z(S \to B \to S') \times z(S' \to B' \to S'')$. In this way, the mental model can be used to estimate the probability of any sequence of behaviors leading from one state to another. Namely, the probability that using a sequence of behaviors $B_1, \ldots, B_l$ is successful in leading from an initial stimulus S_1 to a final stimulus S_l is estimated as

$$\Pr(Q) = \prod_{j=1}^{l-1} z(S_i \to B_i \to S_{i+1}) \tag{11.2}$$

where Q is a shorthand for the sequence

$$Q: \quad S_1 \to B_1 \to \cdots \to B_l \to S_l \tag{11.3}$$

What stimuli should be considered as goals to reach? It is natural to consider stimuli with primary value, such as food or water. However, one can also consider learned goal stimuli. For example, a student might learn that completing homework is a goal (see also sections 12.2.2 and 15.4.4).

11.4.3 Decision Making

Decision making in a thinking mechanism can be subdivided into three steps:

1. Choose a set of possible sequences to evaluate.
2. Estimate the value of each sequence.
3. Use estimated sequence values to select a behavior.

In general, these steps must be repeated every time a decision has to be made, that is, before choosing each behavior. This is necessary because the outcome of a behavior may not be the one that was planned for. Among the three steps listed, the first one is the most difficult, because there may be thousands of potential courses of action even in relatively simple environments. We consider this point last. Let's imagine for now that we have a number of possible sequences to choose from. For example, these may be alternative routes or other behavioral sequences leading to the same food source, or sequences leading to alternative food sources. The second step, calculating the value of each sequence, can be accomplished using probability formulas like equation 11.2 to gauge which sequences are likely to succeed. For example, if only the last stimulus in a sequence has value, the expected reward from pursuing a sequence Q that leads to S_l is

$$v(Q) = u(S_l)\,\Pr(Q)$$

where Pr(Q), the probability that the sequence succeeds, is calculated as in equation 11.2. This basic formula can be modified to take into account that sequence value may be affected by the cost of behaviors, that many stimuli in a sequence may have value, and that sequences may have different lengths. For example, the following formula takes into account both the cost $c(B_i)$ of behavior B_i and the value of all stimuli in the sequence:

$$v(Q) = \sum_{i=1}^{l} [u(S_i) - c(B_i)] \prod_{j=1}^{i} z(S_{i-1} \to B_i \to S_i)$$

Once a value has been calculated for all sequences under consideration, we are ready for the third and final step: using sequence values to choose a behavior. As mentioned above, it is generally not optimal just to choose the behavior with the highest expected value, because decision making should also leave room for exploration. In our chaining model, we solved the problem using the decision-making formula in equation 3.8. This formula preferentially chooses behaviors with higher value, but also chooses lower-valued behaviors some of the time, with the amount of exploration regulated by the parameter β. We use the same strategy in our thinking model, by assigning to behavior B a value that equals the expected value of all sequences that start with B, or 0 if none of the evaluated sequences starts with B. In this way, even behaviors that are not currently known to lead to any reward will be explored some of the time. In addition to satisfying the requirement for exploration, using the same decision-making equation in chaining and thinking makes the two mechanisms more easily comparable. The only difference is that thinking assigns values to behaviors using the mental model to look for profitable behavioral sequences, while chaining uses incrementally learned stimulus-response values.

Let's now go back to the first point in our decision-making process: How can a thinking mechanism decide which sequence to evaluate? An ideal mechanism would consider all possible sequences, but this is only realistic in very simple environments. A laboratory rat, for example, may be described effectively as choosing between pressing a lever and doing a few other things (often bundled together as "not pressing"), but a rat in a natural environment may have hundreds of potential food locations to travel between. Picking promising sequences among myriad possible ones is a very complex problem, and potentially very time consuming (Sutton and Barto 2018). Here we adopt the

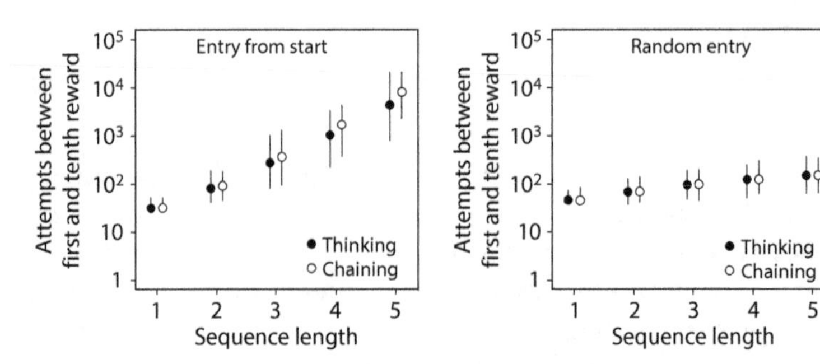

FIGURE 11.2. Comparison of thinking and chaining in learning sequences of actions to obtain a reward. Learning speed is measured as the number of attempts between collecting the first and tenth rewards. (There is no difference between the two mechanisms until the first reward is collected; after the tenth reward, learning is practically complete.) Left: The sequence is always entered from the first step. Right: The sequence is entered from a random step (see section 3.5). Bars show the full range of values over 500 simulation runs. See section 1.5.5 for simulation details.

brute-force strategy of considering all paths that can be inferred from the current mental model, starting from the current stimulus and leading to all known valuable stimuli. This is adequate because we explore relatively simple problems and use a computer, but in general a thinking mechanism must have a strategy to select which sequences to consider for evaluation. We return to this point in section 11.7.

11.5 Comparison of Thinking and Chaining

We are now ready to compare the efficiency of thinking and chaining. We study examples representative of the problems discussed in section 11.3, starting with learning a sequence of actions leading to a goal, S_{goal}. We consider this problem in two variants: when the sequence is always entered from the first step and when it is entered from a random step (section 3.5). In both cases, we assume that a mistake leads back to the start of the sequence. As we discussed in section 3.8, many problems encountered in nature have a similar structure.

The results of our simulations are in figure 11.2. We argued earlier that thinking has no advantage over chaining when learning single responses. The simulations confirm this conclusion, which we can also reach mathematically as follows. When S_{goal} is reached for the first time, thinking will start learning that $S \to B \to S_{\text{goal}}$ is a possible transition, where B is the correct behavior and S the starting situation. Every time B is chosen and the transition is experienced, the estimated value of the sequence $Q: S \to B \to S_{\text{goal}}$ increases as follows:

$$\begin{aligned}\Delta v(Q) &= u(S_{\text{goal}})\Delta z(S \to B \to S_{\text{goal}}) \\ &= \alpha_z u(S_{\text{goal}})\left[1 - z(S \to B \to S_{\text{goal}})\right]\end{aligned} \tag{11.4}$$

using equations 11.1 and 11.2. In the case of chaining, the same experience increases the stimulus-response value $v(S \to B)$ as follows (assuming the learned stimulus value of S_{goal} is 0):

$$\begin{aligned}\Delta v(S \to B) &= \alpha_v\left[u(S_{\text{goal}}) - v(S \to B)\right] \\ &= \alpha_v u(S_{\text{goal}})\left[1 - \frac{v(S \to B)}{u(S')}\right]\end{aligned} \tag{11.5}$$

We now note that equations 11.4 and 11.5 are equivalent, because z ranges from 0 to 1 while $v(S \to B)$ ranges from 0 to $u(S')$. Thus, if $\alpha_z = \alpha_v$, thinking and chaining will learn the correct response at the same speed.

With longer sequences, the simulations in figure 11.2 show that thinking has an advantage. The advantage grows with sequence length, because a longer sequence means a longer stretch of unrewarded experiences before the reward is found for the first time. During this time, chaining learns nothing while thinking learns a mental model of the possible sequence steps, which can then be used to plan the sequence. However, the advantage of thinking is surprisingly modest. It can be meaningful for longer sequences that are always entered from the start, but it appears minute for short sequences and for sequences that can be entered from any step, as in these cases chaining is also very efficient. True random-entry sequences are uncommon in the real world, but many sequences are entered from states at various distances from the goal, which can greatly shorten the learning time for chaining (section 3.8).

Thinking offers clearer benefits when it can use a mental model that has already been learned. For example, a territory may have multiple sources of food, such as fruit trees, which can be productive or unproductive at different times. Chaining learns little from visiting an unproductive tree, while thinking can learn a mental model that quickly leads back to the tree once it starts to bear fruit (this scenario resembles our initial example in section 2.5). In this and similar cases, the left panel of figure 11.3 shows that thinking has a sizable advantage over chaining when a known food source becomes unproductive and a new one becomes productive. The right panel of figure 11.3 shows the advantage of thinking in a related situation, that is, when the environment is explored for a while before any rewards are introduced. As we saw in chapter 7, this design was used in experiments on latent learning in rats,

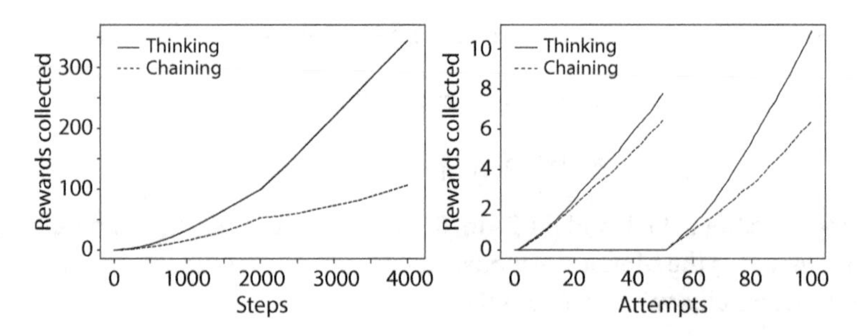

FIGURE 11.3. Comparison of thinking and chaining when a mental model remains useful after an environmental change. These simulations take place in a 4 × 4 environment representing a territory where an organism may forage. The organism leaves from location (1,1) for foraging trips and returns to this location after it finds food. Left: A food source is available at location (4,4) until step 2000, after which it moves to location (4,1). Thinking learns the sequence from (1,1) to (4,4) a bit faster than chaining (compare with figure 11.2), but it has an even bigger advantage after the food changes location. Right: The two lines at the left replicate the difference between thinking and chaining in learning the sequence from (1,1) to (4,4). The two lines at the right refer to learning the same sequence, but after spending some time in the territory without finding any rewards. In this case, thinking displays a larger advantage because it can learn a mental model of the environment even when there are no rewards.

although the results of those experiments agree more with chaining than with thinking.

The advantages of using mental models are even greater when the models can be built and modified quickly, as it can be accomplished with human language. For example, you may hear from a friend of a road closure on your way to work. You can then integrate this information with your knowledge of the road network to devise an alternative route to work, without ever experiencing the road closure directly. As we discuss in later chapters, and especially in chapter 15, the quick acquisition of mental models for various tasks is a crucial ingredient in the success of human thinking.

11.6 Costs of Thinking

In the preceding section, we showed that the advantages of thinking are most pronounced when learning longer sequences of actions (especially those that are always entered from the start) and when a previously acquired mental model is relevant for the current task. So far, however, we have neglected a number of important difficulties that any thinking mechanism faces.

11.6.1 Slow Decision Making

Slow decision making could jeopardize any advantages of thinking over associative learning. We have already acknowledged this difficulty above, when we pointed out that exploring all possible sequences of actions before making a decision would take too much time in any realistic situation. We discuss this point in more detail in the next section, but we can already note that human thinking is, in fact, generally slow compared to simple decision making (section 7.7). For example, chess players, from novices to experts, can analyze only about four moves per minute (Gobet and Charness 2006). This speed is clearly too slow for many everyday tasks faced by animals and humans alike, such as deciding in which direction to flee an approaching predator or which berry to pick next. Thus thinking can only be beneficial when sufficient time is available for decision making.

11.6.2 Complexity of Mental Machinery

Using a mental model for thinking requires rather complex mental machinery. Both long- and short-term memory must be able to accurately encode sequences of arbitrary events, which is not typical of animal memory (section 5.4.2). Furthermore, sequences must be recalled and manipulated accurately at the time of decision making. The costs associated with this mental machinery are hard to quantify precisely, but they may represent an important evolutionary bottleneck because chaining incurs very few of them. In chaining, neither decision making nor long-term memory needs to handle sequential information, and the requirement on short-term memory is just to remember which stimulus S and behavior B precede a valuable stimulus S'. Moreover, chaining requires a smaller memory. For example, with n stimuli and m behaviors, chaining requires storage for at most $n \times m$ stimulus-response values plus n stimulus values, while a mental model storing $S \to B \to S'$ triplets requires up to $n^2 \times m$ storage locations. Finally, we should note that an emerging thinking mechanism is not going to be perfect at first, and an imperfect mechanism may have more difficulties competing with chaining.

11.6.3 Stimulus Representation

In this book, we have not discussed stimulus representation in depth: we have simply assumed that mental mechanisms receive information about which stimuli are present. In reality, nervous systems laboriously extract stimulus information from raw sensory data (sections 1.3.3 and 3.2). It is worth considering this process now, as it poses some difficulties to thinking mechanisms that associative learning largely forgoes.

Consider what stimulus perception is for, say, a hen. Instead of neat labels, such as "fox," "bug," "chick," and so on, the hen's brain receives unlabeled patterns of millions of neural signals, which it must somehow identify. The diversity of sensory experience is such that, most likely, each pattern is seen just once in a lifetime. Yet, the hen successfully identifies the vast majority of the stimuli it encounters. Simplifying a bit, we can say that it does so by exploiting similarities within and differences between classes of stimuli (Blough 1990, Arbib 2003, Ghirlanda and Enquist 2003, Koenen et al. 2016). For example, stimuli to be understood as "chick" have features in common that tell them apart from "fox" stimuli. The goal of sensory processing is to extract these features and make them available for learning and decision making. A stimulus S becomes thus an array of feature values $s_1, s_2, \ldots, s_n$. Extracting useful features is far from simple, but it is relatively well understood and is often modeled successfully using artificial neural networks (Arbib 2003, Enquist and Ghirlanda 2005).

Stimulus representations with many features are easily accommodated in associative learning. For example, in chapter 6 we modeled social learning by considering that each stimulus can have "social" and "nonsocial" features, each of which can acquire stimulus-response and stimulus values. This approach can be used with any number of features and has been successfully applied to natural as well as artificial intelligence (Arbib 2003, Enquist and Ghirlanda 2005, Sutton and Barto 2018), including the A-learning model (Enquist et al. 2016). In thinking mechanisms, however, stimulus representations must support mental sequencing, in addition to encoding relevant features. For example, our model performs mental sequencing by looking for $S \to B \to S'$ triplets that have identical start and end stimuli. However, with a realistic, multidimensional stimulus representation, stimuli will be more (or less) similar to each other, rather than being identical (or not). Suppose an organism has witnessed 10 experiences of the triplet $S_1 \to B_1 \to S_2$ and 10 of $S_2 \to B_2 \to S_{\text{reward}}$. So far, we have assumed that each triplet has just one representation memory, in which case it would be relatively easy to conclude that they can be sequenced. However, a more realistic assumption is that each experience gives rise to different representations of S_1 and S_2, and hence to a unique triplet. This introduces two complications. First, the mechanism now needs to decide whether the start and end stimuli of triplets are "similar enough" that their sequencing is worth considering. Second, looking for triplets that can be sequenced will involve considering 100 different pairs of triplets, instead of just 1. More generally, if each triplet has q representations in memory, the number of sequences to consider when planning a sequence of l actions will increase exponentially by a factor q^l. This is yet another combinatorial dilemma, which, unless it is handled cleverly, would decrease the speed of thinking.

In summary, when stimuli are represented realistically in terms of multiple features, thinking mechanisms require additional complexity compared to associative learning. We argue in the following that these problems can be addressed by symbolic representations, which, however, begs the question of the origin of these representations and of the mental operations that are performed on them.

11.7 The Mental Library

Natural selection can favor the evolution of thinking only when productive behavioral sequences can be found through thinking more efficiently than through associative learning. Furthermore, the benefits thus gained must offset the costs of more complex mental machinery. Our analysis suggests that the conditions that favor thinking may be more rare that one might guess. One reason is that chaining is surprisingly efficient in learning short sequences (and even longer ones when entry patterns are favorable), and it is economical in terms of mental machinery. In this section, we begin to explore how thinking can become more successful, and thus how it may evolve. This exploration continues through the end of the book.

The main idea is that thinking becomes more useful when it is provided with more information—either about the world, in the form of an effective mental model, or about how to make the thinking process itself more effective. We can imagine that this information is stored in a "mental library" that the thinking process can consult to speed up the search for productive sequences. In this section, we consider what kind of information the mental library can hold, what advantages and disadvantages it conveys, and where the information comes from.

11.7.1 What Is in the Mental Library?

To illustrate how a mental library can help thinking, we use examples from the history of computer chess, and then we generalize to thinking in general. The history of computer chess itself parallels the history of artificial intelligence (section 9.5.2; see also Marsland 1990, Nilsson 2009).

The rules of chess are deterministic and relatively few, such that a mental model for playing chess is relatively easy to construct (we do not consider the problem of learning the rules). However, Shannon (1950) pointed out early on that the sheer number of possible move sequences is so large that mental simulation would need an effectively infinite time to discover the best move in the vast majority of board positions. In fact, move sequences are subject to the same kind of combinatorial explosion that we have been discussing since chapter 1. Even if there are just 3 possible moves in each position, simulating all possible 40-move games (about the average length of a game of chess) would entail simulating $3^{40} \simeq \frac{1}{2}10^{40}$ moves for each player (5,000 billion billion billion billion). Several strategies have been used to circumvent this difficulty.

One strategy is to rely on precomputed databases of best moves for some positions. One such database, produced using a supercomputer, enables perfect play from any position with up to seven pieces on the board (Zakharov et al. 2019). Consulting a database is not "thinking" per se, but it can aid thinking because it helps with evaluating move sequences not in the database. For example, if there are eight pieces on the board, a chess program needs only to simulate move sequences until it generates all accessible seven-piece positions, at which point it can use the database to evaluate candidate moves. Similarly, any thinking mechanism can use known partial solutions to lighten the thinking load. For example, if you are trying to get home from an unfamiliar part of town, but you know how to get home from Central Square, you only need to figure out how to get to Central Square. Another example is memorizing multiplication tables of single-digit numbers, which helps us calculate products of larger numbers.

Another strategy is to have procedures to solve particular sets of problems, such as how to win when playing with a king and a rook against a king alone or in other simple scenarios (Marsland 1990, Nunn and Burgess 2009). These procedures are very common in human thinking. For example, we can readily multiply a base-10 number by 10 by adding a zero at the end, which gives, in a shorter time, the same result as carrying out the multiplication. It is a major part of cognitive psychology to investigate how these procedures are learned and used (Anderson 1982, 1993).

Finally, crucial advances in computer chess have resulted from finding ways to focus the search process on more promising sequences. This has involved leveraging heuristics used by human experts (e.g., trying to occupy the center of the board) and advances in search algorithms. For example, it is sometimes possible for an algorithm to determine that a given move will always lead to worse outcomes than another move. At that point, all sequences starting with the bad move can be ignored. Even with these strategies, however, chess programs still had to evaluate tens of millions of moves to beat the best human players. Recently, search has been improved further by training artificial neural networks to estimate the value of board positions, such that moves leading to lower-value positions can be discarded (Lai 2015, Silver et al. 2018). In this way, chess programs can find excellent moves by simulating only about 10,000 moves per decision. This technique is interesting because it is closer to how human chess players think, that is, recognizing good and bad positions more than consciously searching many move sequences (Gobet and Charness 2006). More generally, human experts in many domains rely on recognizing patterns that are meaningless to novices (Chi et al. 2014, Ericsson et al. 2018). Pattern recognition can also be thought of as a procedure in the mental library, although often individuals cannot fully verbalize how it works.

11.7.2 How Large Is the Mental Library?

The size of an individual's mental library can be staggering. Skilled chess players have been estimated to know the value of more than 100,000 positions, which they can use to quickly find good moves in other positions (Gobet and Charness 2006), while experts in mental arithmetic may know hundreds of tricks to solve specific problems (Benjamin and Shermer 2006). These tricks are not always hard but are typically unknown to novices. For example, the squares of numbers close to 50 can be calculated easily through the formula $(50 + x)^2 = 2500 + 100x + x^2$, which involves an easy multiplication by 100, the squaring of a small integer (likely already known), and adding the fixed number 2500. Thus, to compute 54^2, we start with $54 = 50 + 4$ and get $2500 + 4 \times 100 + 4^2 = 2916$. Efficient tricks can also involve unconventional manipulations, such as treating numbers as text. For example, to square numbers ending in 5, one can write the number as $x5$ (where x is all the digits apart from the final 5), then calculate $x(x + 1)$, and finally append 25 to the result. Thus we can calculate $85^2 = 7225$ by just performing $8 \times 9 = 72$. This trick works because $x5$ represents $10x + 5$, so that the square is $(10x + 5)^2 = 100x^2 + 100x + 25 = 100x(x + 1) + 25$. Because of place value, by appending 25 to $x(x + 1)$, we simultaneously multiply the latter by 100 and add 25.

We have drawn examples from chess and mental arithmetic because they are thoroughly studied, not because we think they are exceptional in any respect (Rieznik et al. 2017). As we saw in chapter 9, humans pursue a myriad of activities, such as jobs, college degrees, and hobbies, and each of these is supported by a large number of learned facts and mental skills. This diversity stands in stark contrast to associative learning, which consists of just a few fixed processes.

11.7.3 How Useful Is the Mental Library?

In this section, we attempt to quantify how the mental library can improve the efficiency of thinking. For the sake of brevity, but also for generality, we do not discuss in detail how the mental library is used in any concrete case. Rather, we simply assume that it enables us to rapidly evaluate some sequences of actions, so that the thinking mechanism needs to evaluate fewer sequences. For example, an expert chess player uses her mental library to immediately discard many bad moves, after which she evaluates only relatively few move sequences (Gobet and Charness 2006).

As a simple model of the effects of a mental library, consider a thinking mechanism that has time T available for decision making. In this time, it needs to evaluate sequences of l actions, each action being drawn from m possible ones. (The model in this section is very similar to those in sections 1.4 and 4.4, but here we count sequences evaluated mentally rather than behavioral sequences.) Ideally, the mechanism would consider all m^l possible sequences, but the time T is generally not enough. If the mechanism can evaluate one action per unit of time, or one sequence per l units of time, the percentage of sequences that can be evaluated is

$$\%\ \text{evaluated sequences} = 100 \times \min\left\{\frac{T}{lm^l}, 1\right\} \tag{11.6}$$

where the min operation ensures that the result is at most 100% (the time T is sufficient to evaluate all sequences). Let's now introduce the mental library. We assume that it enables the thinking mechanism to evaluate immediately, on average, a fraction s of the m possible actions, at each step of a sequence. For instance, in a chess example, the mental library would enable discarding the worst sm moves at each step in a sequence of moves. This amounts to knowing right away the value of $s^l m^l$ sequences, or a fraction s^l of possible sequences. After using the mental library, the thinking mechanism proceeds as above, evaluating as many as possible of the remaining $(1-s)^l m^l$ sequences. Thus equation 11.6 is replaced by

$$\%\ \text{evaluated sequences} = 100 \times \min\left\{s^l + \frac{T}{l(1-s)^l m^l}, 1\right\} \tag{11.7}$$

which is graphed in figure 11.4. The figure shows that the mental library can be very effective, but, at least in our simple model, for longer sequences its effect is sizable only for relatively large s. This suggests that a relatively large mental library may be required to significantly increase the efficiency of thinking about longer sequences.

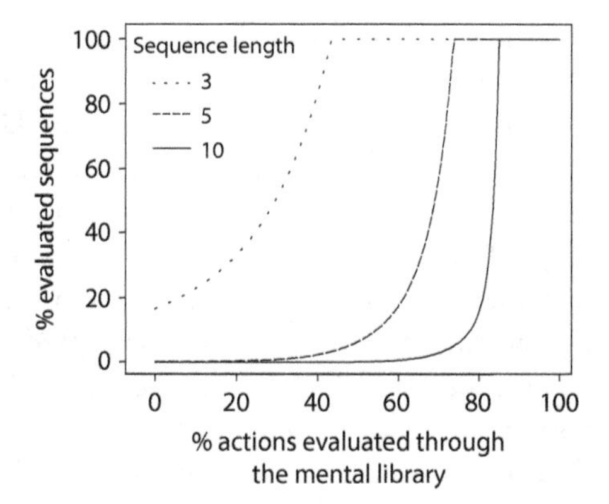

FIGURE 11.4. Percentage of possible sequences that a thinking mechanism can evaluate, as a function of the fraction of actions s that can be evaluated using a mental library at each step in a sequence. The curves graph equation 11.7 with $T = 100$, $m = 5$, and $l = 3, 5, 10$.

11.7.4 How Costly Is the Mental Library?

A mental library has costs as well as benefits. A large memory is an obvious requirement for a large mental library. Moreover, similar to mental models, mental skills that include multiple steps require accurate representation of sequential information. In domains as diverse as mental arithmetic, tool making, music playing, and cooking, the various steps dictated by a procedure must typically be followed in strict order.

We also note that items in the mental library rarely provide direct answers to problems. For example, knowledge that "dense objects sink" is not enough to enact productive behavior. The relevance of this information to the current situation must be recognized and integrated with other available information. In other cases, more than one item in the mental library may be applicable. For example, 55^2 can be calculated with a general-purpose multiplication algorithm, or with either of the two shortcuts in section 11.7.2. Thus an effective thinking mechanism needs not only the ability to reason about sequences and a large mental library but also the ability to integrate these two components.

11.8 Cultural Evolution of Thinking

One of our hypotheses about human evolution is that uniquely human mental skills are mainly a product of cultural evolution (section 2.4). We have already argued for this hypothesis in the previous two chapters, based on the empirical evidence that uniquely human skills require lengthy training and extensive social input, and on the fact that many of these skills are too recent to be a direct outcome of genetic evolution. This chapter offers a more theoretical argument in support of the hypothesis. By comparing thinking and associative learning, we see that thinking is more efficient than associative learning only in certain circumstances, primarily when learning longer sequences of actions. In these cases, however, finding good sequences among the myriad imaginable ones can take too much time. For thinking to become viable, the ability to plan using a mental model must be complemented by a large mental library that speeds up decision making by helping the thinking mechanism evaluate potential sequences. Furthermore, any circumstance that speeds up the acquisition of the mental model and the mental library themselves would also improve the efficiency of thinking.

The connection of these results with our hypothesis is that we suggest that mental models and the mental library are primarily of cultural origin. This is both because they pertain to skills of recent origin and because an individual could not discover them on its own to any meaningful extent. Importantly, we believe that, in a strong sense, the ability to think is itself largely a cultural product. In addition to learning many facts and solutions to specific problems that have a long cultural history, we also learn mental models of physical and social reality (e.g., number concepts and social institutions), as well as general mental skills that can themselves be considered part of the thinking process, including skills to effectively use mental models and the mental library. The next two chapters explore how humans can learn a large number of behavioral and mental skills.

12

Acquisition and Transmission of Sequential Information

- Efficient information transfer between individuals requires accurate processing of sequential information.
- Social transmission of information enables individuals to learn about behavioral sequences, causal relationships, and episodic information.
- Learning from others may include observations of sequences of behaviors and sequences of subgoals.
- The social transmission of sequential information greatly aids the transmission of culture.
- Thinking and social learning can amplify each other, providing a powerful machinery for acquiring information, making inferences, and acting.
- Symbolic communication (language) endows humans with even more efficient and versatile information transmission.

12.1 Introduction

As discussed in the previous chapters, humans appear to have uniquely developed abilities to receive and store sequential information. This ability has far-reaching consequences for cultural evolution because much cultural information is sequential in nature and cannot be transmitted efficiently, if at all, as individual elements. Imagine, for example, learning a musical piece by being told the notes but not their order, or figuring out the content of this book by knowing the words but not the sequence in which they appear. Indeed, even knowing all ordered pairs or triplets of notes and words would not be very helpful. The sequential structure of information in human mental life is extensive.

We summarized some domains in which sequential information is crucial in table 9.1. Without faithful transmission of sequential information, either through direct imitation, language, or other symbolic codes, the opportunity for cultural evolution in all these domains would be much more limited (Bandura 1986, Tomasello, Kruger, et al. 1993, Sterelny 2003, Lewis and Laland 2012, Heyes 2018, Gelman and Roberts 2017, Papa et al. 2020). The ability to receive and store sequential information is also necessary for thinking. In the previous chapter, we highlighted how thinking requires mental models that encode information about the environment, as well as mental skills that can use the mental model to reach better decisions. The mental model itself consists of sequential information about cause-effect relationships, and mental skills for thinking must also be capable of handling sequential information.

In this chapter, we consider in some detail the costs and benefits of various ways of transmitting sequential information. These include social learning of behavioral sequences from observation, learning about sequential episodic and semantic information, transmission of sequential information through language, and teaching. In the most sophisticated case, sequential information can be transmitted precisely through a symbolic code (a language), even outside of its applied context. For example, we may convey verbally how to cook a certain dish without actually performing the cooking. The chapter ends with a brief discussion of the role of cultural evolution in shaping the acquisition and transmission of sequential information, and in the use of such information, which we expand upon in the following chapters.

12.2 Transmission of Behavioral Sequences

In this section, we consider various mechanisms through which individuals can learn behavioral sequences by observing others. Consider an individual that observes another performing the following sequence:

$$S_1 \to B_1 \to ... \to S_n \to B_n \to S_{\text{reward}} \tag{12.1}$$

This sequence is perceived by the observer as a sequence of stimuli (section 6.3) from which various kinds of information can be extracted. For example, just paying attention to the occurrence of rewards can be informative and can result in simple forms of social learning (chapter 6). Here we consider learning that uses more detailed information about the observed sequence. Further information about behavioral sequence learning can be found in Heyes (2013) and Loucks et al. (2017).

12.2.1 Imitation of Behavioral Sequences

A capacity to imitate behavioral sequences means that the observer can remember and reproduce at least some of the observed behaviors *in the*

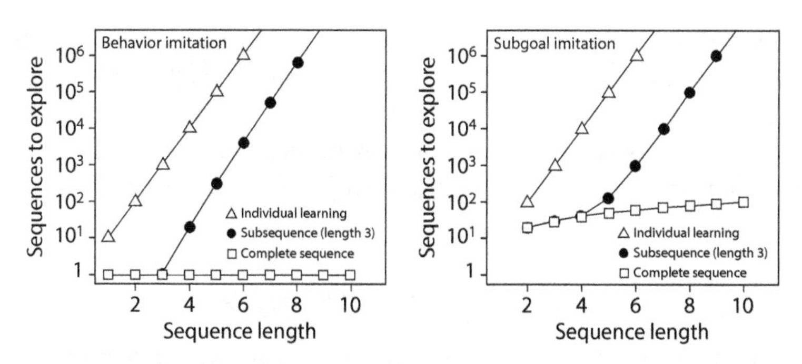

FIGURE 12.1. The number of sequences that an individual can explore, under different assumptions about what the individual can learn from social observations. Being able to explore more sequences offers more opportunities and flexibility, but the learning cost quickly rises. Left: The consequences of being able to observe and imitate others' behavior; see equation 12.2. Right: The consequences of being able to observe and remember subgoals, including their order; see equation 12.3. In both panels, the behavioral repertoire size is $m = 20$.

observed order. To illustrate the advantage of behavioral sequence imitation over individual learning quantitatively, we consider an individual trying to learn a sequence of l behaviors, each chosen out of m possible ones. As discussed in chapter 1, learning such a sequence by random exploration is practically impossible unless m and l are very small, because the expected number of attempts required to perform the correct sequence by chance is equal to the number of possible sequences, m^l. The sequence can be learned more rapidly, however, if the individual can observe others performing the sequence. For example, suppose the individual can remember a subsequence of $b \geq 1$ behaviors in the observed sequence. Then the number of remaining sequences to attempt is reduced to

$$(l - b + 1)m^{l-b} \tag{12.2}$$

where the first factor derives from the assumption that the individual does not remember *where* the remembered sequence fits into the full sequence. In the best case, that is, when the full sequence is remembered ($b = l$), the number of attempts is just 1, that is, the sequence is learned right away. Any $b \geq 1$, however, is better than not remembering anything. The left panel in figure 12.1 illustrates this result. Unless the behavioral repertoire is very small, social information can decrease the learning time dramatically.

There are also costs associated with the imitation of behavioral sequences. To be effective, social learning by direct observation requires more sophisticated cognitive abilities than standard associative learning and chaining. First, the individual must know how to imitate behavior, that is, how to translate an observation of a behavior into the corresponding behavior (chapter 6; see also Heyes 2018). Second, the individual must be able to remember the order of the behaviors it observes. These requirements become particularly demanding when motor flexibility is high, as each behavior in the sequence can come from a larger set.

12.2.2 *Imitation of Subgoals*

An alternative to remembering behaviors in an observed sequence is to remember their effects. That is, in the behavior sequence in equation 12.1, the individual would remember the sequence of stimuli that appear and then try to find behaviors that reproduce these stimuli. In this sense, the stimuli can serve as subgoals on the way to the reward. If all stimuli are remembered in the correct order, the number of sequences to explore would be just ml, because each step in the sequence can be explored independently (we assume for simplicity that all stimuli arising in the sequence are distinct). Thus the effect on the number learning time is dramatic: an exponential growth, m^l, versus a linear growth, ml, of the number of sequences that needs to be explored. Learning time will be considerably shortened even if the individual can remember only some of the stimuli, for example, a sequence of k stimuli. In this case, the number of sequences to explore are

$$mk + m^{l-k} \tag{12.3}$$

Note that in a sequence of length l there are $l-1$ possible stimuli that can serve as subgoals and thus that $0 < k \leq l-1$. The right panel of figure 12.1 shows that remembering stimulus sequences and using them as subgoals can yield substantial advantages, even though behavioral imitation seems somewhat more efficient. For example, remembering all behaviors enables one to immediately learn the full sequence, while remembering all stimuli still leaves ml possible sequences to explore. However, using subgoals also makes it easier to know when a behavior fails to have the intended outcome, which may be important—especially when action outcomes are not deterministic.

The costs associated with imitating subgoals are similar to the costs of imitating behaviors. Subgoal imitation is somewhat simpler, as it does not require translating observations of behaviors into one's own behaviors, but it still requires mental machinery to faithfully represent and process sequential information. Additionally, the strategy of imitating subgoals must somehow be known to the individual.

In summary, imitation of behavioral sequences and subgoal sequences can be highly effective, alone or in combination. However, this conclusion is subject to two caveats. First, imitation skills must be reasonably accurate. For example, systematic errors in remembering sequences may make social learning even less efficient than individual learning. Second, there must be sufficient information in the social environment to offset the cost of the additional mental machinery for sequential imitation.

12.3 Semantic and Episodic Memory

In addition to information about behavioral sequences, humans also learn sequential information about specific events and about general properties of the world. Memory of specific events that an individual has experienced is called episodic memory, while knowledge about the world accumulated through life is referred to as semantic memory (Tulving 1972, 2002). Sequential information is prominent in both, ranging all the way from memories of daily events, such as who arrived first to a party, to general knowledge, such as the order of seasons or that causes must precede effects. According to our hypotheses, both episodic and semantic memory rely on a domain-general, inborn ability to process sequential information (chapter 10).

There is ongoing discussion about what is stored in episodic memories and about how semantic and episodic memories relate to each other (Greenberg and Verfaellie 2010, Michaelian 2016, Mahr and Csibra 2018, Renoult et al. 2019, Quiroga 2020). Also debated are the evolutionary benefits of episodic memory and whether it is unique to humans (Suddendorf and Corballis 1997, 2007, Schacter et al. 2011, Tulving 2005, Boyer and Wertsch 2009, Allen and Fortin 2013, Michaelian 2016, Mahr and Csibra 2018). In this section, we touch upon some of these issues as we consider the role of semantic and episodic memories in thinking.

12.3.1 Sequential Aspects of Semantic and Episodic Memory

Broadly construed, semantic memory comprises any information about the world that is not about a specific event, such as information about useful behavioral sequences and about cause-effect relationships. Both kinds of information are based on elementary sequences of the kind "using behavior B in situation S_1 causes situation S_2," or $S_1 \to B \to S_2$. These elementary sequences can be arranged in longer sequences, as well as in networks (such as the representation of a subway system) and in hierarchical structures (such as a tree of possible move sequences in chess). As we discussed in the previous chapter, these elementary sequences can be used for planning and decision making. Sequential structures are also prominent in models of thinking from cognitive psychology, such as semantic networks (Sowa 1987, 2014).

Semantic networks are based on "semantic triples" that are structurally identical to $S_1 \rightarrow B \rightarrow S_2$ triples, but in which S_1 and S_2 refer to two concepts, and B to a relationship between them. For example, two possible semantic triples are

$$\textit{Bird} \rightarrow \text{Has} \rightarrow \textit{Wings}$$

$$\textit{Wings} \rightarrow \text{Enable} \rightarrow \textit{Flight}$$

Semantic triples also have sequential structure, as most relationships are not symmetrical. For example, "*Wings* → Have → *Bird*" makes no sense. Similar to the use of $S_1 \rightarrow B \rightarrow S_2$ triples in planning, semantic networks can be used to reason about the concepts and relationships that are represented. For example, sequencing the triples listed above would enable one to conclude that birds can fly because they have wings. (There are exceptions to such conclusions, if the premises are not absolutely true.)

Turning to episodic memory, its most intriguing features are that episodic memories are continuously and spontaneously formed, even of seemingly trivial facts, such as the hairstyle of a post office worker or what shoes you wore at your last birthday. It is likely that these features are inborn. Episodic memories are often ordered (we know, or can infer, which of two events in memory occurred earlier), and contain within them sequential information about how an event developed in time, such as the order of songs heard at a concert. However, episodic memories also tend to fade. We remember many things that occurred today, fewer from yesterday, and many fewer from last month. Details within each memory are also forgotten over time, and may be "filled in" using semantic knowledge. For example, you don't need to remember that you unlocked the car before entering it, because that is what usually happens.

12.3.2 Semantic and Episodic Sequences in Thinking

The main features of episodic and semantic memory fit well with the hypothesis that these memories have evolved to provide complementary kinds of information for thinking. First of all, it seems likely that mental simulation and planning are primarily grounded in semantic rather than episodic memory (for alternative views, see Suddendorf and Corballis 1997, 2007, Atance and Meltzoff 2005, Michaelian 2016, Schacter and Addis 2007). In fact, semantic memory holds the kind of information about the world that one typically needs in planning. In addition, it would often be too time consuming to sift through all relevant episodic memories when making a decision. For example, we may have thousands of experiences of shopping at our regular grocery store. It seems far more efficient to summarize these experiences in semantic knowledge, such as "the vegetables are near the entrance," rather than to go through all memories of the grocery store when we need vegetables.

Nevertheless, episodic memories can aid thinking in many ways, arising from the fact that information that is not useful in the present may become useful in the future (as we saw in section 2.5). Thus episodic information can be used to update or complement semantic knowledge (Klein et al. 2009, Allen and Fortin 2013, Mahr and Csibra 2018). For example, we might notice a special vegetables sale in a different section of the grocery store, or that the vegetables have been moved. We might not be shopping for vegetables when we experience these facts, but it will be useful to remember them for future shopping trips. This kind of episodic memory is likely to be very useful for everyday planning, and is also likely to be short-lived. For example, it may not matter *right now* that the car keys are in the bathroom rather than in their usual place, but it will probably matter soon enough. On the other hand, it matters little where the keys were last week.

To understand the role of semantic and episodic sequences in human mental life, we should also note that both kinds of sequences can be communicated between people, which greatly magnifies the possibility for teaching, cooperation, and cultural evolution in general. These topics are covered in the next three chapters.

12.4 Language and Other Symbolic Codes

> But above all wondrous inventions, how great was that of who imagined how to communicate his most hidden thoughts to any other person, even if greatly removed in time and space? To speak with those who are in the Indies, to speak with those who are not yet born, and who will not be for a thousand or ten thousand years? And how easily? With the various combinations of twenty little characters on paper.
>
> —GALILEO GALILEI, *Dialogue on the two chief systems of the world* (1632)

In humans, social transmission of sequential information often occurs through symbolic codes rather than direct observation. Spoken language, the only such code for most of human history, was eventually complemented by other symbolic systems, such as writing, mathematics, maps and graphs, or musical notation. Even though the mental mechanisms behind symbolic codes are still debated, these codes are clearly fundamental to human cultural evolution and other uniquely human mental and behavioral phenomena (Jablonka and Lamb 2005). In fact, symbolic codes are used both to store and mentally process sequential information and to transmit information between individuals.

Transmission of sequential information can benefit from symbolic codes in at least two ways. First, symbolic codes can bypass the need to observe the behavior or situation to be learned about, making learning possible at any time and place. Second, symbolic codes can transmit information compactly, hence at a faster rate. To appreciate these advantages over social learning by direct observation, consider the following thought experiment involving a computer task. In each trial, m shapes appear on-screen, and the participant can score a point by touching l of them in a predetermined sequence. This is a simple implementation of our usual sequence learning task, with an expected learning time of m^l trials when learning by random trial and error. In a simple case with 10 shapes and a sequence of 2, the expected learning time is $10^2 = 100$ trials. In section 12.2.1, we showed how learning by observing an experienced individual can shorten this learning time. Using language, the improvement is even more dramatic: it is just the time the experienced individual needs to tell the correct sequence to the learner (plus some time for the learner to actually memorize the sequence). Symbolic transmission becomes particularly efficient when the task itself is time consuming or requires effort to demonstrate.

Symbolic codes also expand the range of what can be transmitted, such as information about the environment, including causal relationships and past events, and information about how to think, speak, or perform other mental tasks. Consider, for example, this simple sentence: "To get to the gas station, turn right after the red house and continue straight; however, if you get to the supermarket, you went too far and you must turn around." This sentence conveys, in a few seconds, information about how to reach a goal (the gas station), about spatial relationships in the environment, about how to correct a potential mistake, and about a location (the supermarket) that can be recalled to achieve a different goal (shopping) at another time. In summary, the potential advantages of symbol-based social learning are numerous. We explore some of them in the upcoming chapters; we now turn to some requirements and potential disadvantages.

For symbolic transmission to be possible, symbolic codes must have been learned previously, and learning codes and translations between different codes is time consuming. For example, it takes years to master a language. Even if language has specific genetic support, at least the lexicon is purely of cultural origin and has to be learned without prior information. Another drawback with symbolic codes is that they can easily convey false information. It is just as easy to say "the world is round" as "the world is flat." This problem is much less severe when learning is based on direct observation of reality. However, the expressive freedom of language enables us to imagine what does not exist. This is ultimately the source of many human cultural phenomena, some of which are explored in chapter 15.

12.5 Teaching

A teacher arranges specific experiences that facilitate learning and provides feedback to ensure that learning is successful. As a form of social learning, teaching deserves special consideration because it can greatly increase learning speed, compared to situations in which the learner simply observes or interacts with individuals who do not attempt to facilitate learning. That teaching is vital to human development has been recognized many times (Bruner 1966, Vygotsky 1978, Premack and Premack 1996, Sterelny 2003, Bjorklund and Causey 2017, Hattie and Clarke 2018, Lucas et al. 2020). We considered the efficiency of teaching when we described animal training in chapter 3. There we saw that, by arranging appropriate experiences, the time to learn a sequence of length l can be made linear in l, rather than exponential as in random trial-and-error learning. The same teaching techniques apply to human learning, but they can be made even more efficient by the use of symbolic codes, such as language and writing, as discussed above. Symbolic codes also help the learner to actively engage the teacher ("I don't understand this step; can you help me?").

We remark that teaching is particularly important for learning sequences of mental operations (e.g., mental arithmetic), since these are hard to infer from observations of behavior. This concept becomes highly significant in chapters 13 and 15, in which we consider the social transmission and cultural evolution of mental processes.

12.6 Cultural Evolution of Sequential Information Processing

Because sequential structure is crucial to most human cultural phenomena (chapter 9), faithful encoding and transmission of sequential information seem necessary for cultural evolution on a significant scale. What is the origin of the mental skills that make such transmission possible? While we believe that humans have unique inborn abilities for sequential information processing (chapter 10), we also think that many uniquely human mental skills are themselves the product of cultural evolution and are learned by individuals. We thus envision a feedback process in which cultural evolution refines sequence transmission skills that in turn make cultural evolution a more powerful creative process. In our view, human sequence learning is thus a flexible and dynamic ability that is largely learned socially. This view contrasts with the idea that memory and learning abilities are predominantly inborn.

Learned sequential information of the kind discussed in this chapter becomes particularly useful when combined with mental skills, such as thinking, which can make productive use of such information. It should be clear that social learning can enhance both: observation and symbolic transmission of information can efficiently provide information about the world, and can transmit mental processes that improve thinking ability, such as methods of planning, deduction, and inference. In chapter 13, we consider how mental skills can be socially transmitted, while chapter 15 is dedicated to the interplay of social learning and thinking in cultural evolution.

13

Social Transmission of Mental Skills

- Uniquely human mental abilities comprise a diversity of phenomena, including thought processes, fast sequence-to-sequence translations, and fast learning of complex information.
- Many uniquely human mental skills are behaviorally silent and thus cannot be learned by observation.
- However, verbal interactions between active learners and tutors can transmit a multitude of mental skills present in the social environment.
- Individuals can acquire increasingly complex mental skills through codevelopment of mental skills and language.
- Empirical data on childhood development are consistent with the hypothesis that human mental skills are socially transmitted.
- The social transmission of mental skills sets the stage for their cultural evolution, including the cultural evolution of language skills, learning skills, and teaching skills.

13.1 Introduction

In this chapter, we explore our third major hypothesis about human evolution: that uniquely human mental skills, such as thinking, planning, and remembering, are mainly learned, whereas how animals learn and decide is essentially hardwired (chapter 2).

How can humans learn mental skills? In this chapter, we survey the characteristics of human mental skills, and leverage findings from developmental psychology to argue that the transmission of mental skills is ubiquitous in human development. We then examine the human ability to transmit mental skills socially, in light of the fact that mental processes cannot be directly observed. We propose that humans solve this difficult problem with a variety of approaches, including a significant time investment, the active role of tutors and learners, the widespread use of symbolic language, and the codevelopment of mental skills—that is, the process whereby emerging mental skills support each other's development. Although we believe that mental skills for sequential information processing have a genetic basis that is unique to humans, we also discuss the extent to which learning is necessary for these skills to attain their full potential.

Our overall goal in this chapter is to stress the centrality of development for human mental skills and cultural evolution. On the one hand, developmental processes determine how mental and behavioral skills are transmitted through social learning. On the other hand, cultural evolution can itself shape the mental skills that are transmitted, in a feedback process that we discuss more thoroughly in chapters 15 and 16.

13.2 Characteristics of Human Mental Skills

A mental skill may be defined as a sequence of mental processes that transform information from an initial to a final state. To illustrate some characteristics of mental skill transmission in humans, we start with an example in which a teacher transmits a relatively simple skill: how to calculate the volume V of a cone or pyramid. Provided that the learner is sufficiently experienced, the teacher may simply state that V can be obtained by multiplying the base area b by the height h, and then dividing by 3 ($V = bh/3$). After this, when asked to calculate V with a base area of 1 and a height of 2, the learner can correctly answer 2/3. A number of tests of this kind can demonstrate that the skill has been transmitted correctly. This apparently mundane example can be unpacked to demonstrate many remarkable features of mental skill transmission.

We note first that the teacher uses a symbolic code to express the relationship between volume, base area, and height as the formula $V = bh/3$. Indeed, the teacher may use a combination of several codes, such as spoken and written language and mathematical notation. Symbolic codes can also transmit smaller bits of information, such as $\pi = 3.14\ldots$ or "Paris is the capital of France." However, symbolic codes carry a large initial learning cost, which is necessary for mutual understanding between teacher and learner (section 12.4). Thus, to understand the formula $V = bh/3$, the learner must understand the concepts of volume, area, height, multiplication, division, and equality and how these are represented symbolically within the formula. In general, transmission of a mental skill is possible only if the the learner has sufficient preliminary knowledge. If the groundwork has been done, however, mental skill transmission can be fast and accurate. The learner only needs several exposures to the formula $V = bh/3$, after which it can be applied to all possible cones. This efficiency and accuracy stands in stark contrast with the painstaking nature of trial-and-error learning.

Another feature of mental skill transmission is that it involves sequences of effortful, conscious operations (Johnson-Laird 2006, Kahneman 2011). Our example features only a short sequence (a multiplication and a division), but, given enough time and opportunity for practice, it is possible to transmit much more complex sequences, such as lengthy food preparations, the construction of complex artifacts, like a kayak or loom, or techniques to solve differential equations.

With the notation introduced in section 10.3, we can represent mental skills as sequences of the form

$$S \rightarrow P_1 \rightarrow M_1 \rightarrow P_2 \rightarrow M_2 \rightarrow \ldots \rightarrow M_n \rightarrow P_n \rightarrow B \qquad (13.1)$$

where S is an initial stimulus, P_i are mental processes, M_i intermediate memory states, and B a behavioral output. S and B may be missing, if there is no external trigger or behavioral output, or they may themselves be sequences, such as the stimuli generated by reading a written question and the movements necessary to write down the answer. For example, S may be "What is 102×6?" and B the writing down of 612. The Ps would then be intermediate steps, such as 100×6, 2×6, and $600 + 12$, with the Ms corresponding to the results of these computations. Additionally, several Ps would be needed to plan the sequence of operations, such as deciding to split 102 into 100 and 2, then multiply each by 6, and finally sum the two partial results. Different individuals may go through different sequences of steps when executing complex skills.

With enough training, humans can also learn transformations between different kinds of sequences, which eventually are performed with great accuracy and little apparent effort. This ability is demonstrated routinely in translating between different languages, in reading, in which sequences of visual inputs are transformed into linguistic sequences, and in handwriting and typing, which translate linguistic sequences into sequences of movements. Other examples are the translation of written music into music "in one's head" or into the movements necessary to perform it on an instrument, and of heard music into a symbolic representation as a sequence of notes. Similarly, a skilled computer programmer can translate an abstract algorithm into concrete implementations in one or more programming languages. Such feats require a flexible ability to process and transform mental sequences. This flexibility is essential to both the diversity of learned skills and the frequent use of acquired skills to cope with unforeseen circumstances. For example, an experienced commuter can integrate information about a subway stop closure with knowledge of the subway system to plan an alternative route to work. Utilizing mental skills in everyday, small-scale planning is a hallmark of human mental activity, as one day's activities are seldom identical to those of the day before.

In conclusion, human mental skills are characterized by a large toolbox of mental procedures, or *Ps*, that are acquired over the course of a lifetime (section 11.7). These make up the bulk of an individual's mental abilities, enabling us to carry out information-processing steps, like computing intermediate results, making plans, and reorganizing information. Thus human mental flexibility stands in stark contrast with our conclusion that animal intelligence may rely upon a general inborn machinery of associative learning (chapter 8). We also note that sometimes human mental activity is only loosely coupled to environmental stimuli, such as when a writer plans a novel or an architect plans a building. Below we focus on social transmission of *Ps* for the storage, recall, and processing of information. We pay less attention to classic questions, such as whether learning is implicit or explicit, which memory systems are involved, and the exact content of each *P*, although these are also important for a complete understanding of human cognition.

13.3 How Can Mental Skills Be Socially Transmitted?

In contrast with behavioral skills, mental processes cannot be directly observed: there is nothing obvious for the learner to copy, and no direct way for the teacher to observe what has been learned. These difficulties appear so daunting that, in many domains, uniquely human mental skills have often been assumed to have a strong genetic component. The most famous example is that of language (Pinker 1994), but many other skills have been suggested to be largely innate, including numerical abilities (Butterworth et al. 2018), imitative skills (Meltzoff et al. 2018), and other social skills (Tooby and Cosmides 1992). However, empirical evidence about the development of these skills is most compatible with the suggestions of authors like Vygotsky (1978), Piaget (1977), Bruner (1966), Jablonka and Lamb (2005), Hockett (1960), and Heyes (2018), who see a preponderant role of social transmission in the development of human mental skills. In this section, we summarize this evidence, considering mental skill acquisition to be the product of the following four elements:

1. A substantial time investment in mental development.
2. The active role of both learners and tutors in the transmission process.
3. The use of language and other symbolic codes to establish mental links between individuals.
4. The codevelopment of language, internal symbolic representations, and mental skills.

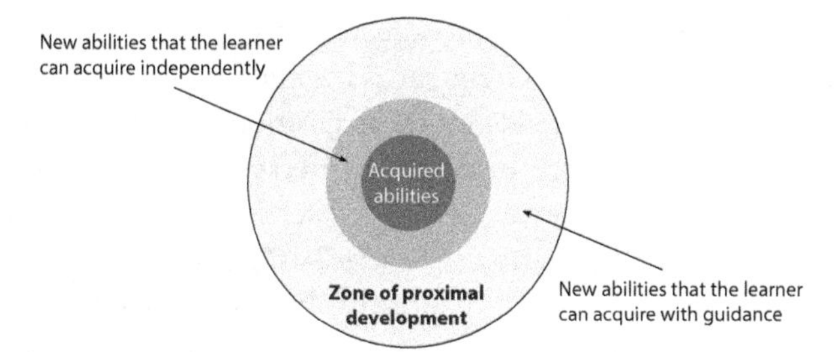

FIGURE 13.1. Vygotsky's "zone of proximal development" illustrates how tutoring helps learners master skills they could not acquire on their own, provided the skill is within reach (Vygotsky 1962).

The first point appears uncontroversial: humans in all societies need to spend years and even decades learning in order to become fully competent individuals. Below we discuss the remaining three points.

13.3.1 Active Tutors and Active Learners

It is well established that parents and other tutors play an important role in the acquisition of uniquely human abilities (Bruner 1966, Vygotsky 1978, Pressley and Woloshyn 1995, Premack and Premack 1996, Tomasello 2014, Bjorklund and Causey 2017, Hattie and Clarke 2018). Vygotsky's "zone of proximal development" is a classic illustration of the role of tutors (figure 13.1). At every stage of development, there are skills that learners can master without help and skills they cannot master regardless of how much help they receive. The zone of proximal development lies in between, and consists of what learners can master only with the help of a tutor. As more skills are mastered, the zone shifts toward more difficult skills.

The role of tutoring is most often studied in educational settings, where much evidence demonstrates the importance of using organized, appropriate teaching methods, communicating learning goals, providing timely feedback, and adapting the material to the learner's level (Pressley and Woloshyn 1995, Hattie 2008, Hattie and Clarke 2018). Perhaps more crucial to our argument, however, is the evidence that active tutoring is omnipresent even outside of educational institutions, and is fundamental for the acquisition of basic mental skills in infancy and early childhood (Rogoff 1990, Bjorklund and Causey 2017). For example, Conner et al. (1997) found that toddlers learned to stack blocks more efficiently when parents provided specific instructions ("Put that block there" or "That one is too big") than when parents either engaged little or demonstrated. Heckhausen (1987) provides another example of active tutoring in an experiment that followed mothers with children aged 14 to 22 months. Using tasks in which blocks had to be fitted to slots,

Heckhausen showed that the mother's feedback adapted to the child's level, becoming more verbal and less frequent as the child became more proficient. The importance of parental guidance for acquisition of mental skills has been demonstrated in many domains, including literacy, numerical abilities, and planning (Tamis-LeMonda et al. 1992, Bornstein et al. 1992, Charman et al. 2000, Morales et al. 2000, Hattie and Clarke 2018, Gauvain et al. 2018).

Because mental skills cannot be directly observed and are time consuming to learn, their social transmission also requires motivated learners who actively seek information (Vygotsky 1962, Bruner 1966, Karmiloff Smith 1992, VanLehn 1996, Winsler et al. 2009, Bjorklund and Causey 2017). Indeed, learning efforts also proceed in the absence of tutors (Flavell 1976, Wolcott 1991, VanLehn 1996, Ericsson 2006, Hattie and Clarke 2018). The importance of individual effort grows with age as the child gains a better understanding of how to learn and develops the ability to monitor her own learning (Brinums et al. 2018). For experienced learners to succeed, it is often enough that the tutor presents learning goals and educational materials. Active learning also requires the learner to correct what she already knows when experiencing limitations and contradictions (Festinger 1957, Piaget 1977, Siegler 1996, 2006). Learning mental arithmetic, for example, depends crucially on private mental practice (Ashcraft and Fierman 1982, Siegler 1996), because unobservable skills are reconstructed rather than directly copied (Acerbi and Mesoudi 2015). Thus the details of mental processes can differ across individuals with similar skill levels. Lemaire and Arnaud (2008) identified as many as nine different strategies for mental addition of two-digit numbers and demonstrated individual differences in their use. Some developmental psychologists deem self-motivated learning so important that they characterize children as "naive scientists" who continuously formulate and test hypotheses about their environment (Gopnik et al. 1997, 1999, Bjorklund and Causey 2017).

In summary, we can identify several feedback loops in the interactions between learners and tutors that greatly facilitate the transmission of mental skills (figure 13.2; Hattie and Clarke 2018). In the first loop, the tutor can provide information about a skill and the learner can provide feedback about her own understanding, either explicitly by asking questions or implicitly through her own performance. This tight interaction between learners and tutors appears to be a uniquely human trait, absent even in our closest relatives (Premack and Premack 1996). A second loop involves the learner and her interaction with the external environment, providing the opportunity to evaluate whether a skill has been mastered. For example, a student may check solutions to problems, or a child may observe whether her block construction stands or falls. Finally, the learner can monitor and evaluate her

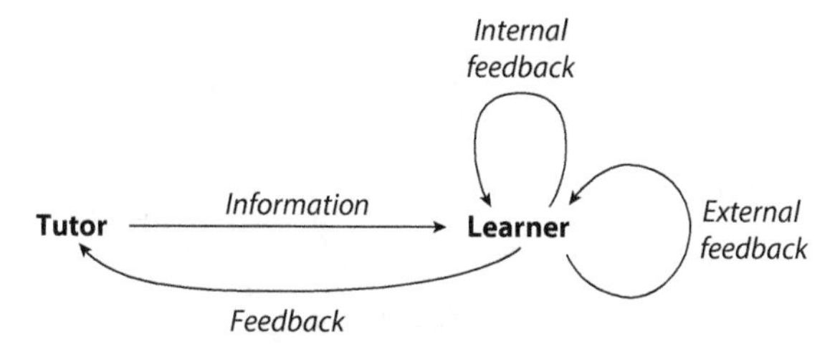

FIGURE 13.2. Feedback loops in human social learning. Internal and external feedback refer to self-monitoring of progress. The former could involve trying to fulfill some mental processing goal, and the latter trying to fulfill some goal within the environment. Feedback from the learner to the tutor represents observable events, such as task performance and requests for help, that the tutor can use to aid the learner.

own mental processes and attempt to improve them. For example, an elementary school child might discover that it is more efficient to compute $54 + 23$ as $54 + 20 + 3$ than to count 23 units starting from 54. The productive operation of these feedback loops requires a considerable time investment and the active participation of both learner and tutor. It also requires a means of communication between the two, which we discuss next.

13.3.2 *Behavioral Silence and Language*

Because mental skills cannot be directly observed like behavioral skills, language is involved in almost all transmission of mental skills. For example, consider a mental process P that modifies a memory from M to M':

$$M \rightarrow P \rightarrow M'$$

All of M, P, and M' are internal to the individual, having no behavioral expressions. The only feasible solution to convey how P works is that the tutor makes it "visible" through a language code. The active social interaction of language enables bidirectional communication between tutor and learner (figure 13.3). For much of human history, spoken language was the only available code. It was supplemented just a few thousand years ago by writing, and then by formal languages in logic and mathematics. This account of the role of language in mental skill transmission is consistent with contemporary theories of development that emerged from the pioneering work of Vygotsky and Piaget (Bruner 1966, Halliday 1973, Piaget 1977, Vygotsky 1978, Nelson 1998), which stresses how language, symbolic coding, and joint attention are used to establish "mental links" between individuals (Moore and Dunham 1995, Tomasello 2003, 2010, Jablonka and Lamb 2005, Heyes 2018).

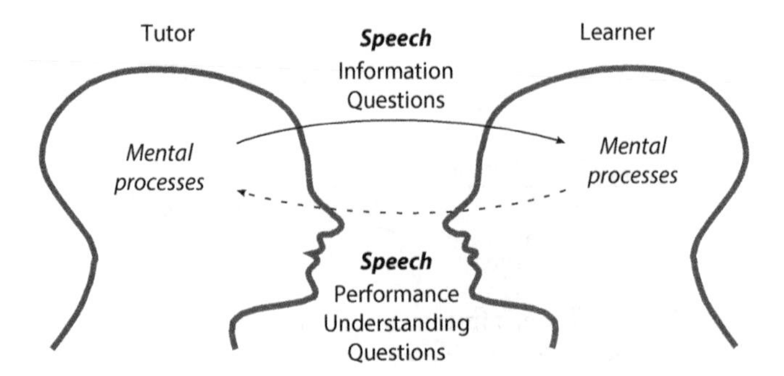

FIGURE 13.3. Language and other symbolic communication can transmit behaviorally silent information. This process includes the tutor making mental information available to the learner, and the learner informing the tutor about what it has or has not learned, in order to facilitate feedback. (Graphics from openclipart.org.)

The role of language is best appreciated by trying to imagine doing without it. In this case, the transmission of mental information would have to rest on inferences from observations of stimuli and behavior. For example, you may observe stimulus S and behavior B but not the intervening mental processes:

$$S \to (P_1 \to M_1 \to P_2 \to M_2 \to P_3) \to B$$

In general, it is impossible to deduce the underlying thought processes through simple observation of a series of stimuli and behavior. There are formal proofs of this statement (Luenberger 1979, Hitchcock 2018), but perhaps the best intuitive argument is that even our most precise language can convey only part of what goes on in the tutor's mind. The mental processes that escape linguistic expression are so hard to decipher that they have kept philosophers and scientists busy for thousands of years. Indeed, every advancement in our understanding of the human mind has required painstaking collection of vast amounts of data, as well as logical and scientific rigor. On the other hand, even complex information, when it is clearly understood by the tutor, can be communicated relatively easily through language.

For example, consider a child who has learned a few number words, say, 1, 2, and 3, but she has not yet learned their ordering or that they signify magnitudes. From this starting point, the child can be taught to say the numbers in sequence, even without reference to cardinality. The tutor can then convey the meaning of numbers by establishing, for example, a finger-pointing counting process to identify sets of 1, 2, or 3 toys. Through repetition, the tutor can ensure that the child uses numbers in the correct sequence and that the counting process is used correctly. The process is necessarily slow because the child neither knows the order of 1, 2, and 3 nor has a concept of counting. The correct order and the counting procedure are transmitted through physical and verbal interaction with the tutor, and the end result is that the child's mind contains mental structures that are functionally similar to the tutor's. Language can be used during this process not only to repeat the words for 1, 2, and 3 but also to provide feedback, such as "Good job" or "No, 3 comes *after* 2," and to direct the interaction through prompts, such as "Repeat with me" or "Try on your own." The child can also use language herself to ask for help or feedback. Note that a productive interaction requires the child to have some initial proficiency in language. Indeed, for effective mental skill transmission, language cannot be limited to labels for concrete or even abstract objects, such as a chair or the number 1. A much wider set of symbolic representations is needed, including labels for internal mental processes ("recall," "add," "imagine"), categories of stimuli ("toys," "numbers"), and sequential operations ("before," "after," "first"). These symbolic representations are themselves acquired gradually, building on what the learner already knows.

13.3.3 Codevelopment

Crucial to the interplay between cognition, culture, and development is that the acquisition of a given mental skill depends on previously established mental skills. A complicating issue is that language and other symbolic communication systems are themselves mental skills that require development. For instance, everyday skills that we consider basic, such as counting to 10 or making a shopping list, are assembled from more elementary skills that took a long time to learn as children (Carey 2009, Bjorklund and Causey 2017). Exploring and understanding the social transmission of mental skills is made more challenging by their interdependence: we cannot neatly separate the skills from the experiences they rely on. For example, when learning to count, a child may learn some number words, something about the concept of ordered sequences, how many teddy bears she owns, and something about the meanings of "larger" and "smaller." Empirically, we see that when mental skills build on each other, they are acquired in a typical order, but for skills that depend less on each other, the order of acquisition is less rigid (Berk 2013). For instance, number words must be learned before using them to count, and number concepts before arithmetic operations. In contrast, one child may learn to recite the numbers 1 to 10 before gaining the ability to count objects, whereas another may learn just the sequence from 1 to 3, apply it to counting, and then learn the sequence from 4 to 10. A third child may learn first to recognize the written symbols for 1 and 2, and so on.

Although codevelopment makes research on mental skills acquisition harder, we believe that the interdependence of mental skills presents a dual opportunity for social transmission. First, building mental skills from the bottom up allows tutors to check that learning is proceeding correctly each step of the way. For example, the tutor can verify that a correct counting procedure has been learned before introducing addition and subtraction. Second, the interdependence of different mental skills may increase the accuracy of social transmission, because mistakes in skill acquisition tend to be evident when practicing or learning other skills. We do not mean to imply that mental skill transmission has become trivial for humans—it is still laborious. Rather, we argue that codevelopment, the use of language, active teaching and learning, and a considerable time investment make mental skill transmission possible for humans, while it is impossible, or nearly so, for animals. This ability paves the way for the cultural evolution of mental skills that we discuss in the following chapters.

13.4 Development of Sequential Information Processing

Uniquely human mental abilities require skills to perceive, process, and produce sequences (table 9.1). This has led us to hypothesize that humans have a genetic advantage in sequential information processing (chapter 2). At the same time, the evidence summarized above and in chapter 9 shows that human sequential abilities require extensive time and social interaction to develop fully. It is thus plausible that many mental skills for sequential information processing are socially transmitted, even though they build on a uniquely human genetic foundation. In this section, we first sketch a rough timeline for the development of some sequential abilities and then review the evidence that these abilities are socially transmitted. Studies in developmental psychology are rarely designed to investigate basic sequential abilities. However, the tasks employed in this research, which may focus on representation of events in memory, planning, causal understanding, and problem solving, have sequential components.

Sequence abilities are rudimentary in early childhood (table 13.1), but improve steadily, as documented in studies of how event representations develop and are used in planning (Cowan and Hulmes 1997, Carey 2009, Nelson 2009, Bjorklund and Causey 2017, Loucks and Price 2019). For example, when asked to describe a shopping trip, a three-year-old might answer, "I just buy things to eat. We get a cart or a box to hold it. When we're done, we just get in the car and go home" (Hudson et al. 2014, 83). Such a report represents the typical sequence of events in a shopping trip, often referred to as a script. At this stage, however, it is neither a plan of how to shop

Table 13.1. Typical ages for development of some abilities requiring sequential information processing

Ability	Age	Sources
Deferred imitation of action sequences	End of year 1	Bauer et al. (2000), Sundqvist et al. (2016)
Counting	Around 2 years: a few numbers in order	Wynn (1992), Carey (2009), Dehaene (2009)
Speech perception	1.5–2 years	Jusczyk (1997), Gertner et al. (2006)
Language production	2–3 years	Clark (1995), Tomasello (2007), Hoff (2020)
Memory span for digits	2 years: 2–3 digits; 5 years: 4–5 digits	Dempster (1981)
Script, event representation, and episodic memory	1–2 years: some ability; 4 years: extensive ability	Fivush and Hudson (1990), Nelson (1998)
Music	3–4 years	Brandt et al. (2012)
Causal understanding	2–3 years	Gopnik et al. (2001)
Plans and planning	From age 3–5 on	Hudson et al. (1995), Friedman and Scholnick (1997)
Reading and writing	From age 4 on	Chall (1979)

nor a faithful account of a specific shopping trip (Nelson and Gruendel 1981, Slackman et al. 1986, Hudson and Mayhew 2011). Faithful episodic and autobiographic memory develop somewhat later (Bjorklund and Causey 2017), as does the ability to make simple plans (McCormack and Atance 2011, Hudson et al. 2014). The ability to mentally modify sequences also starts to appear in early childhood. For example, according to Hudson and Fivush (1991; see also McCormack and Atance 2011, Hudson et al. 2014), one of the first steps toward general planning abilities is the introduction of "open slots" in scripts that can be filled with different information, depending on need. For example, a shopping script may have open slots for the items to buy or which stores to visit. More advanced abilities to manipulate sequences for planning also develop gradually, including the ability to mentally compare different sequences (Kaller et al. 2008, McCormack and Atance 2011), to partition a task into subgoals, and to make a plan for each (Loucks et al. 2017).

Overall, these findings point to a progressive refinement of sequential abilities during development, such that memories for plans, events, and causal relationships can eventually be effectively distinguished, in parallel with development of the mental skills required to use multiple types of information. However, observations that chart the gradual development of sequential abilities are not sufficient to establish that they are learned socially; this could instead be the outcome of a gene-driven maturational process or of individual learning without significant social input. Circumstantial evidence, however, suggests that active social interactions, as discussed in the previous section, are instrumental to the development of sequential abilities. For example, if these social interactions had no effect on learning, it would not make sense for adults and children to be motivated to engage in them. Furthermore, children go through various revisions of their mental strategies, which would be puzzling if these strategies were genetically coded. Confirming these intuitions, data support the idea that children improve their sequential abilities through interactions with knowledgeable tutors. First, interactions with tutors expose children to countless examples of sequences, their importance, and how to use them: the tutor can recount past events, demonstrate planning by thinking aloud, and prompt the child to remember and reorganize sequential information (Catmur et al. 2009, Haden and Ornstein 2009, Fivush 2014, Bjorklund and Causey 2017, Gauvain et al. 2018). Several researchers have also argued that it is only through discussion of nonfactual events with adults and acquisition of temporal language that children begin to grasp important temporal and sequential concepts, and indeed become capable of mental sequence abilities (Gauvain and Rogoff 1989, Gauvain et al. 2018, Radziszewska and Rogoff 1991, Nelson 1998, Benson 1997, Gauvain and Huard 1999, Hudson 2002, 2006). Second, improvements in sequential abilities have been concretely measured in many experiments. For example, Wood et al. (1978) gave block construction puzzles to three- and four-year-olds, and showed that the children improved the most when their mothers provided feedback that was attuned to the child's actions, rather than when the mothers either demonstrated the whole task or provided scripted feedback.

With the acquisition of language and symbolic representations, it also becomes possible to learn domain-general-sequence processing skills that apply to any kind of sequence. For example, languages contain many elements that apply to any situation in which sequential order matters, such as "before," "after," "first," "second," "start," "end." Languages also make it possible to specify sequential operations, such as "append," "repeat," "reverse," "sort," "remove." This conceptual toolbox enables humans to efficiently convey information about sequences, such as "First, place the nut on the anvil stone, then strike it with the hammerstone" or "Add Lisa to the guest list before Peter," which would be extremely difficult to convey without language. These skills provide combinatorial efficiency because they transfer readily across contexts. For example, we can describe list-sorting algorithms that work for any list, regardless of the nature of its elements. In addition to being a powerful practical tool, the existence of such universal language for sequential information processing is theoretically notable because it indicates that the underlying mental skills are domain-general rather than domain-specific.

In conclusion, experimental evidence is consistent with the hypothesis that the mental skills to represent and process sequences can only develop fully through social learning. At the same time, many aspects of this developmental process are poorly understood. Future human developmental work could focus more directly on simple sequential abilities, for example, through the design of tasks that test the ability to discriminate between stimulus sequences, like those used in animal research (section 5.4.2). This approach could serve to identify the ages at which human sequential abilities surpass those of other species, and would also bring into focus particular sequential abilities, which are difficult to isolate from commonly used complex tasks, such as construction puzzles and route planning.

13.5 Cultural Evolution of Mental Skills

Our conclusion in this chapter is that most uniquely human mental abilities can be learned from sociocultural experiences, that is, from active interactions with experienced tutors. Young children acquire mental skills mostly through direct social interactions, while artifacts like books and other media can partly replace tutors for older individuals. As described in section 9.5.3, Heyes (2018) has expressed similar views.

Several of the points we discussed connect with long-standing issues in philosophy, psychology, and artificial intelligence, including theories of mind, mechanisms of mental abilities, and questions about nature and nurture (for an overview, see Rescorla 2017). Cultural evolution adds another layer to these questions, in that the experiences informing an individual's mental development can change significantly over historical time and geographical location (Prinz 2012, Heyes 2018). Apart from basic genetic support for sequential information processing, it is possible that most mental skills are the product of cultural evolution (see also chapters 2 and 15). Under this view, the human mind is subject to "mental cultural evolution," whereby humans invent, over many generations, symbols and procedures that coevolve with each other and with the social and physical environment. For example, cultural evolution offers a nongenetic solution to the problem of categorizing the complex input provided by sense organs, sometimes referred to as the "easy" *symbol grounding* problem (Coradeschi et al. 2013, Cubek et al. 2015, Harnad

2017). Genetic evolution can yield only basic rules in this domain, such as the recognition of simple features of stimuli, like size or color (chapters 4 and 8). Cultural evolution, on the other hand, can result in concepts that capture the structure of the environment at a deeper level, such as force and acceleration. A similar argument holds for mental skills. Genetic evolution has produced a powerful learning mechanism—associative learning—but it is constrained by combinatorial problems (chapter 8). Cultural evolution has produced a vast array of mental skills that allow for the efficient solution of a much larger set of problems. In other words, cultural evolution is a more efficient solution to equipping an organism with mental skills that are useful in the world, a challenge we call the *process grounding* problem. For example, we have readily available concepts for sequence operations, such as "reverse" or "append," but not for "split a list in the middle and reverse each part" or "reverse the last third of a list and put it at the front." Presumably, the reason is that the former operations are useful tools for thinking and discovering productive behavior, while the latter are useful more seldom. We can also speculate that the mechanisms for cultural transmission, including language itself, have evolved culturally to convey mental information effectively. This is consistent with cultural variation in how children and adults interact, and how children become encultured (Rogoff 1990, Konner et al. 2010, Lew-Levy et al. 2017).

One seeming weakness of the argument that cultural evolution creates uniquely human mental skills is that uniquely human mental skills are themselves necessary for cultural evolution—otherwise, any animal species could engage in cultural evolution on the same scale as humans. As hinted several times in previous chapters (see, e.g., chapters 2 and 10), we believe this conundrum can be resolved if the uniquely human mental skills that enable cultural evolution to bootstrap itself are domain-general abilities for sequential information processing and increased mental flexibility. We consider how these skills may have evolved in the following chapters, especially chapter 16.

14

Cooperation

- Human cooperation surpasses that of other species in its flexibility, diversity, sequence length, and outcomes. It also includes mental cooperation.
- Many benefits of cooperation can be understood in terms of sequences. A group can have more behavioral skills than its individual members, thus enabling the exploration of a larger number of sequences.
- Division of labor reduces combinatorial dilemmas by enabling individuals to work on the tasks they know best.
- Thinking, communication, and negotiation of sequences of actions are crucial elements in most human cooperation.
- Human cooperation relies on the same mental abilities that underlie other uniquely human phenomena.
- Cooperative skills are are subject to cultural evolution.

14.1 Introduction

Along with thinking, use of sequential information, and transmission of mental skills, cooperation is a fourth domain in which humans stand out among animals (Boyd and Richerson 1982, Gintis et al. 2003, Melis and Semmann 2010, Ostrom 2010, Bowles and Gintis 2011, Nowak and Highfield 2011, Sterelny 2012, Sterelny et al. 2013, Levinson and Enfield 2020). Cooperation with others, even strangers, is a feature of practically every domain of human life. Cooperation in animals typically occurs among mates or kin, is often restricted to particular activities or domains, and is supported by strong genetic predispositions (Axelrod 1984, Clutton-Brock 2009). Long, cooperative sequences of behaviors are very rare in animals, and occur mainly in eusocial insects through inborn behaviors, such as in ants and termites (Hölldobler and Wilson 1990). In this chapter, we explore how human cooperation is influenced by the uniquely human mental skills discussed in previous chapters. Section 14.2 shows how cooperation can increase the ability to discover and execute productive behavioral sequences. In section 14.3, we argue that human-level cooperation requires a sophisticated understanding of social relations, or a mental map of the individual's social world. Finally, section 14.4 discusses differences between human and animal cooperation.

14.2 The Power of Cooperation

The essence of cooperation is that joining efforts and skills allows individuals to accomplish together what they cannot accomplish alone. Skill acquisition is time consuming, and a group whose members specialize in different skills can be more productive than if everyone tried to learn every skill (Carneiro 2003, Becker and Murphy 1992, Yang 1994). Rephrased in terms of behavioral sequences, cooperation means that a group of individuals can discover and execute more sequences than any group member alone. For example, two individuals could each perform a different part of the following sequence:

$$S_1 \to B_1^{(1)} \to S_2 \to B_2^{(2)} \to S_3 \tag{14.1}$$

where $B_1^{(1)}$ is an action by individual 1 and $B_2^{(2)}$ an action by individual 2. An assembly line operates in this way. Alternatively, individuals can combine their behavior to perform an action, such as when two people move a heavy item, or when one holds a picture frame and the other nails it to the wall. Such a sequence may be represented as

$$S_1 \to B_1^{(1)} B_2^{(2)} \to S_2 \tag{14.2}$$

We can illustrate the potential advantages of cooperation as follows. Let us first consider sequences without combined actions, as in equation 14.1. Consider two individuals who have a repertoire of m behaviors each. If the individual repertoires are not identical, the joint repertoire will be larger than m, say, $m+k$, with $0 \le k \le m$. The two individuals can then form more behavioral sequences than either can alone, namely, $(m+k)^l$ sequences of length l rather than m^l. If the two repertoires have no behavior in common (an idealized case), the pair can execute $2^l m^l$ distinct behavior sequences, or 2^l times what either individual can execute alone. For $z > 2$ individuals, the upper limit for the group is $z^l m^l$ sequences. If combined actions are allowed, such as in equation 14.2, the number of possible sequences will be even higher.

In addition to being able to execute more sequences, a group of cooperating individuals can also execute *longer* sequences. Consider, for example, a very long sequence assembled by selecting randomly from a set of n possible behaviors. Suppose each individual has a repertoire of $m < n$. We first calculate how many steps an individual can execute before it lacks the behavior required to advance to the next step. We then repeat the calculation for a cooperating group. The probability that an individual knows any given step of the sequence is $p = m/n$. The probability that the individual can execute exactly l steps of the sequence is the probability of knowing the first l behaviors, p^l, multiplied by the probability of *not* knowing the next behavior, $1 - p$:

$$\text{Pr(longest sequence is } l \text{ behaviors long)} = p^l(1 - p) \tag{14.3}$$

Using the properties of geometric series, the expected number of steps that the individual can execute is the average value of l according to the probability distribution in equation 14.3:

$$\text{Expected number of steps} = \sum_{l=1}^{\infty} lp^l(1 - p) = \frac{p}{1 - p} \tag{14.4}$$

To calculate the number of steps that a group of z individuals can execute, we need to calculate the probability that the group as a whole knows each given step, which we write as $p(z)$. This probability depends on how individual repertoires are chosen. If each individual knows m randomly picked behaviors, we have

$$p(z) = 1 - (1 - p)^z \tag{14.5}$$

where $(1 - p)^z$ is the probability that no one in the group knows the behavior required at a given step. If individual repertoires do not overlap, on the other hand, we have $p(z) = zp$ as long as $zp < 1$, and 1 otherwise. Finally, to calculate the longest sequence that the group can execute, we substitute $p(z)$ for p in equation 14.4. The dramatic effect of cooperation, even in small groups, is illustrated in figure 14.1. For example, a single individual needs to know 9 out of 10 behaviors to be able to execute a sequence of length 10. If five individuals cooperate, however, each of them needs to know only ~ 4 behaviors, when repertoires are randomly drawn. With nonoverlapping repertoires, for the group to perform a sequence of any length up to 10, it is enough for five individuals to each know 2 behaviors. It is noteworthy that cooperation enables specialization and division of labor, which entails fewer demands on individuals while making the group more capable (Becker and Murphy 1992, Yang 1994, Carneiro 2003).

In human societies, most cooperative sequences include social elements in addition to behaviors directed toward the physical environment. For example,

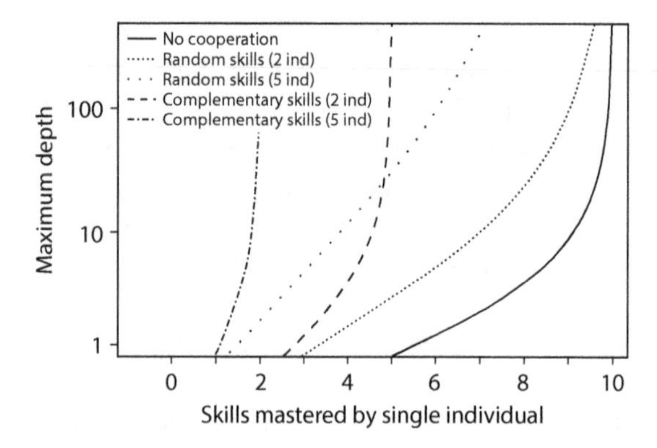

FIGURE 14.1. The expected length of a behavioral sequence that can be executed by combining the skills of cooperating individuals.

humans have organized trade networks that include extensive social interactions. Other organisms exhibit cooperation of this kind, such as nutrient exchange between plants and fungi (Selosse and Rousset 2011), but not on the same scale. The appearance of social elements in productive sequences creates a complex network of interdependence between people and between organizations such as companies and institutions (Wegner 1987). Indeed, the overwhelming majority of both material and nonmaterial cultural products, like bowls or ballet performance, can exist only because of interactions between social entities. Thus, as a consequence of extensive cooperation, humans have become increasingly dependent on one another, a situation that is reminiscent of the evolution of multicellular organisms in which different cell types cooperate (Kroeber 1917, Dobzhansky 1964). We return to this point in chapter 16.

Being capable of executing more sequences is potentially advantageous in the search for productive sequences, but it also incurs time costs (section 4.4) because most potential cooperative sequences will also be unproductive (chapter 1). Together with individual specialization, cultural evolution, and the lifelong acquisition of other mental skills, thinking can help overcome the comparative increase in combinatorial issues for cooperative sequences compared to individual sequences.

14.3 Planning Cooperative Sequences and Cultural Evolution

In this section, we discuss in some detail the steps that go into planning and executing human cooperative sequences, in order to highlight their unique complexity. The first step is to recognize that something of value can be obtained through a cooperative sequence. The complexity of planning the sequence ranges from making small variations in familiar activities, such as when a family plans dinner, to creating entirely novel cooperative projects, for example, a national space program. Cooperation may be based on a single individual making the plan, but it is also common that plans are generated collectively. In the latter case, the planning process itself becomes a cooperative act that takes advantage of the diversity of mental skills, knowledge, and imagination among the individuals (Woolley et al. 2010).

In addition to understanding the physical world, planning for cooperation requires understanding of individual skill sets, motivations, and relationships. Knowledge of individual skill sets ensures that specified tasks within the sequence are performed reliably (Wegner 1987, Lewis 2003). Knowing that Mary is a great plumber, for example, enables allocation of the job and frees the planner from having to understand much about plumbing. Understanding what motivates individuals enables proper incentives and safeguards the cooperative enterprise against cheating and deception (Warneken et al. 2006, Dean et al. 2012). For example, Jane may be motivated by a financial incentive while John by the recognition of colleagues. Similarly, it will be hard to recruit Bob for a job if he does not care about the outcome. Understanding relationships between individuals allows for a productive working environment. Francis may not want to participate in a project with Kim, while Joey and Dave may enjoy working together.

Understanding of the social world is often considered a specialized kind of knowledge that is gathered and processed by inborn mechanisms residing in a dedicated brain module (Cosmides et al. 2010), resulting in "social intelligence" (Jackson 2012) and "theory of mind" (understanding of others; Call 2012). The same outcomes, however, can result from general-purpose mechanisms (Brass and Heyes 2005, Heyes 2012b). For example, a model of the social environment can be learned in the same way as a model of the physical environment, through a combination of direct experience and social learning (Berk 2013). A thinking mechanism, such as the one sketched in chapter 11, could use such a model to plan cooperative sequences, including complex social interactions, thus effectively displaying social competence. There are also mental skills more specific to cooperation. For example, keeping track of who is a good cooperator and who is not requires both an accurate episodic memory and an understanding of others' motives. Both general abilities that promote cooperation and specific skills for cooperation seem to be acquired from an early age (Tomasello 2009, Slocombe and Seed 2019). Such learned skills almost automatically become subject to cultural evolution since they operate directly in the social environment, and uniquely human cooperation and theory of mind may to an important extent be products of such evolution (Heyes 2018).

14.4 Cooperation in Humans and Animals

The contemporary perspective is that humans have a larger inborn drive to cooperate than animals in general (Bowles and Gintis 2011). In addition, humans have established norms and institutions to reward cooperation and trust while punishing noncooperation, so that even less cooperative individuals are encouraged to cooperate (Fehr and Gächter 2000, Richerson et al. 2002, Rothstein and Stolle 2008, Ostrom 2015). These institutions address the basic problem of cooperation—that of free riders who reap benefits without paying costs (Axelrod and Hamilton 1981, Axelrod 1984, Poundstone 1992). For example, tax evaders benefit from government-provided infrastructure without contributing to it. The theory of inborn cooperative drive and social motivation has served to identify many important factors in the evolution of cooperation. However, this does not explain why human cooperation is much more diverse than cooperation of other species, how it can be established quickly and adapted to new circumstances, and why humans seem able to exploit more opportunities for cooperation than other animals.

We propose that the differences in cooperation seen between animals and humans have similar explanations as other differences. Cooperation in animals is based on a combination of genetic predispositions and associative learning. This explains why animal cooperation, which can be impressive, typically is both domain- and species-specific. Examples include cooperation in ants and other eusocial animals, shared parental care, hunting in groups, and mutualistic relationships between species, such as between trees and fungi (Clutton, Brock 1991, Dugatkin 1997, Leigh 2010).

Human cooperation, in contrast, depends on a capacity to effectively plan, communicate, and execute cooperative acts and goals. According to section 14.3, it seems that human cooperative sequences require all the mental abilities that we have found lacking in animals, such as accurate memory for sequences and for episodes of cooperation and defection, a planning mechanism that can operate on complex mental models, and a symbolic language for effective communication of plans and intentions. There are also mental skills specific to cooperation. For example, keeping track of who is a good cooperator and who is not requires both an accurate episodic memory and an understanding of others' motives. Such skills can be learned from an early age and can also be subject to cultural evolution. Indeed, Voorhees et al. (2020) have suggested that both the very ideas that cooperating is good and that noncooperators should be shunned or punished can evolve culturally and be an important driving force for human cooperation.

Note that we do not claim that animals cooperate exclusively when they have strong genetic predispositions for cooperation. Indeed, we can envisage that animal intelligence can result in cooperative behavior without such predispositions. For example, two animals may succeed in obtaining food through a cooperative act and therefore learn to repeat the act (Epstein et al. 1980). However, these occurrences are rare because it is difficult for animals to find such productive combinations of actions, in particular if they require sequential organization.

15

The Power of Cultural Evolution

- Cultural evolution can yield accurate mental models of the world.
- In combination with thinking, cultural evolution can quickly solve problems that are too complex for single individuals.
- Cultural evolution enables individuals to be behaviorally and mentally flexible, because it sidesteps the learning costs of flexibility.
- As a consequence of increased flexibility, cultural evolution can shape many aspects of behavioral and mental mechanisms.

15.1 Introduction

In previous chapters, we suggested that cultural evolution is crucial for understanding the differences between humans and animals. Specifically, we proposed that culture can act as a reservoir for behavioral and mental skills that individuals can acquire much more easily through social transmission than on their own. In this chapter, we show that cultural evolution enables the accumulation of vast amounts of knowledge about the world, much beyond what would be possible for individuals. We also show that the availability of cultural information has momentous consequences for the behavioral and cognitive architecture of individuals, in that the presence of culture allows for increased behavioral and mental flexibility.

15.2 Culture Can Support Learning about the World

In earlier chapters, we discussed various ways in which mental mechanisms more sophisticated than associative learning can result in more efficient learning, such as recombination of sequential information (chapter 11), social learning of behavioral sequences (chapter 12), and advanced cooperation (chapter 14). We have already recognized that each of these is uniquely developed in humans. Here we consider in more detail how cultural evolution can leverage these mechanisms to accumulate knowledge about the world.

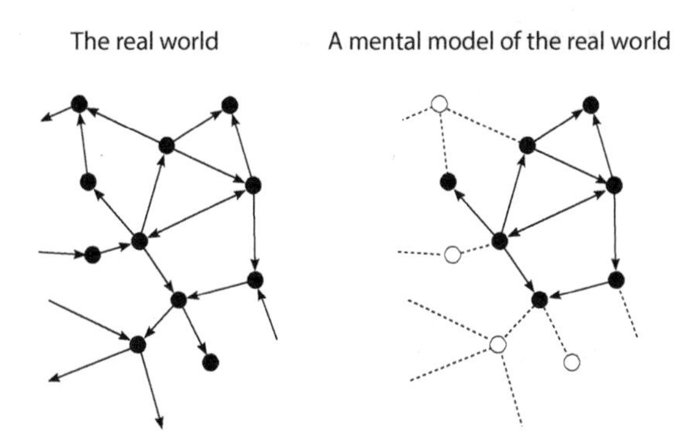

FIGURE 15.1. Left: A world represented as a graph. Each node represents a world state, and arrows represent possible transitions. Each transition is enabled by a given behavior. Right: Representation of an individual's knowledge of the world. Solid circles and arrows represent known world states and transitions; empty circles and dashed lines represent transitions that exist in the world but are unknown to the individual.

15.2.1 Behavior Sequences and Mental Models

In this section, we show how various learning mechanisms can combine to yield cumulative cultural evolution of behavioral sequences. Our main focus is on the amount of knowledge an individual can gather and on how knowledge accumulates over generations. We begin by setting up a model world that contains many behavioral sequences that can be learned.

The model world we consider here is a generalization of the linear sequences of actions we considered in previous chapters. That is, we consider a number of world states, $i = 1, \ldots, N$, as well as the possibility of moving between states through the use of appropriate behaviors. The difference is that states are connected randomly, so that many different paths are possible rather than a single, linear one. As shown in figure 15.1, this model world poses new problems to learners. Is it possible to go from i to j? What is the shortest path? We assume that an individual's task is to solve as many of these problems as possible; that is, to learn the behavioral sequences required to transition between any two world states. We assume that such knowledge is eventually useful, for example, in gathering food, but for simplicity we do not model how knowledge is used.

We consider three learning processes that can be used to gather knowledge about the world. In *individual exploration,* the individual engages in successive learning bouts, each of which starts from a randomly selected state and attempts to discover how to reach other states. This is done simply by selecting a random behavior. If the behavior leads to another node,

Table 15.1. Combinations of learning processes used in the simulations in figure 15.2

Simulation	Individual exploration	Social learning	Recombination of information	Information sharing
Individual learning	100%		No	No
Social learning	75%	25%	No	No
Thinking	100%		Yes	No
Social learning and thinking	75%	25%	Yes	No
Cooperation	75%	25%	Yes	Yes

the transition between the starting node and the new one becomes known. The individual continues to try out random behaviors until it fails to reach another state, at which point the learning bout ends and a new one starts from a randomly selected state. In *social learning,* an individual can explore the world independently as just described, or it can learn from another individual. When an individual engages in social learning, it learns one behavioral sequence, chosen at random from those known to the other individual. Finally, in *thinking,* we assume that the individual effectively stores a mental model of the part of the world that it has explored. The mental model can be used to discover sequences that have not actually been experienced. For example, suppose the individual has never moved directly between i and j, but that it has traveled from i to k and from k to j. Thinking would allow the individual to realize that it can travel from i to j, and the required sequence would be part of its knowledge. An individual that does not think, however, would need to actually experience going from i to j via k for the required sequence to be known.

To explore various scenarios relevant to cultural evolution, we combine the three learning processes as described in table 15.1. All scenarios include individual exploration, since without it nothing can be learned about the world. Social learning and thinking, on the other hand, are present only in some scenarios. In the last scenario, cooperation, we include information sharing, taking into consideration all sequences known to a group of individuals (section 14.2). We simulate discrete generations, and, when social learning is present, we allow the new generation to learn from the old. Other simulation details are in our online repository (section 1.5.5). Dramatically different simulation outcomes are apparent for the first four scenarios, shown in the left panel of figure 15.2. Individual exploration alone results in solving fewer than 10 problems during an individual's lifetime. A small improvement occurs when the learning mechanism is either just thinking (around 25 problems) or just social learning (around 40 problems). In stark contrast, a

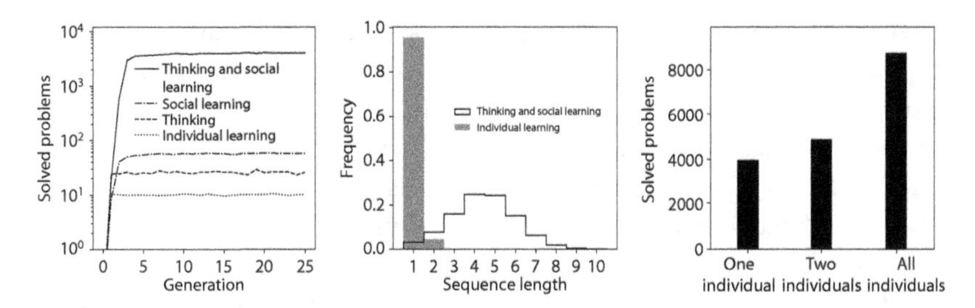

FIGURE 15.2. Simulations of knowledge accumulation in a population, under different combinations of learning processes. Left: Number of problems solved. Middle: Distribution of the complexity of solved problems at generation 25. Right: Benefits of cooperation and joint problem solving comparing one, two, or all individuals together. The model world is a random graph with 100 nodes and ~5% randomly selected direct connections between nodes (figure 15.1). Problem complexity is the number of behaviors required to go from the starting node to the end node. Each individual has 200 learning opportunities in its lifetime, and there are 100 individuals per population. See section 1.5.5 for simulation details.

combination of thinking and social learning results in almost 4000 solutions after just 10 generations of cultural evolution, out of a total of 9900 problems. The middle panel of figure 15.2 shows that cultural evolution based on a combination of thinking and social learning can solve complex problems, while each mechanism on its own mainly learns sequences of one or two steps.

The right panel illustrates the benefits of cooperation and joint problem solving in the presence of both thinking and social learning. After 25 generations, an individual alone can solve around 4000 problems, while two individuals together can solve about 1000 more problems, in part by using longer behavioral sequences. A population of 100 individuals, as a whole, can solve about 9000 problems, or more than 90% of all problems. The benefits of cooperation are even more pronounced in a more complex world and in cases where individuals are allowed to specialize on subsets of problems and share their knowledge to formulate joint solutions (chapter 14).

15.2.2 Productive Representation of the World

In the above examples, we assumed that individuals represented accurately each state of the world (the nodes of the graph in figure 15.1) and only had to learn how to move from one state to another. In reality, representing the environment productively is not easy. For example, a useful representation of the world for a farmer would include the likelihood of rain, but it may not be obvious how to derive this from cues such as cloud features, wind direction, season, and so on. In addition to productive behavioral sequences, cultural evolution can also accumulate knowledge about productive representations of the environment, which we discussed in section 13.5 in terms of symbol grounding. If representations of the world need to be learned, cultural evolution would initially proceed more slowly than in figure 15.2, but it would ultimately produce the same end results through the accumulation of knowledge about world states.

The importance of a productive representation of the world can be illustrated as follows. Assume there are n possible situations an individual can experience. This is a very large number, which must be substantially reduced to avoid combinatorial dilemmas (section 1.3.3). For example, an individual could map each situation to one of k symbolic categories, where $k << n$, and then reason using categories rather than the stimuli themselves. For example, a "storm approaching" category would enable a farmer to plan a course of action regardless of what specific stimulus has been mapped to the category (strong wind, sudden drop in air pressure, information from a nearby village, warning on the radio, etc.). Using categories in this way can substantially reduce the number of nodes in a mental model and thus facilitate planning and execution of behavioral sequences. However, the categories must accurately capture the world, which is known in philosophy as the problem of "natural kinds" (Brzović 2021).

Categorizing a complex world usefully is daunting because of combinatorial explosions. For example, $n = 10$ stimuli can be split into two categories of $k = 5$ stimuli each in 252 ways. With just $n = 50$ stimuli, we can form 126,410,606,437,752 categories of $k = 25$ stimuli. The problem is actually even more complex, because the individual does not know in advance how many useful categories there are or how many stimuli should be in each category. Cultural evolution can solve this problem by refining symbolic systems over generations, retaining the more productive categories devised by individuals (associative learning can tackle this problem nonsymbolically; see sections 5.2 and 11.6.3). Each individual need only generate a small improvement for cultural evolution to result in productive symbol systems. Jablonka and Lamb (2005) have stressed the importance of cultural evolution of symbolic representation and symbolic inheritance systems in human evolution, and consider it an evolutionary dimension on its own.

15.2.3 Culturally Evolving Mental Models

A mental model of the world can be built with symbols that label nodes and actions. If such internal labels can be translated to a common language code, mental models can be shared among individuals (chapter 13) and thus can participate in cultural evolution. The transmission and successive evolution of mental models can be seen as a fairly general model of cultural system evolution (Buskell et al. 2019). Given that sufficient adaptive filtering occurs (section 15.4.3), cultural evolution can result in mental models that support productive interaction with the external world, enabling individuals to make plans that stretch considerably into the future.

These conclusions notwithstanding, the practical benefits of cultural evolution can vary substantially. For example, cultural systems may also incorporate false beliefs and beliefs that are not about reality. Further, culture may be less useful if the environment changes so quickly that existing solutions become obsolete before new ones can be discovered (Boyd and Richerson 1985, Enquist and Ghirlanda 2007). Other factors that influence the outcome of cultural evolution are the efficiency of exploration and of thinking, which affect the rate at which new solutions are discovered (chapters 4 and 11), while the efficiency of social transmission influences how quickly culture can accumulate (chapter 12).

15.3 Culture Can Support Behavioral Flexibility

We observed in chapter 9 that motor flexibility is a key feature of human uniqueness. While most animals have a limited, genetically determined repertoire of motor patterns, human motor skills are primarily learned and extremely varied, including pursuits such as manufacturing stone tools, playing the violin, skiing, or typing 100 words per minute. Such extreme flexibility is puzzling: although a larger behavioral repertoire puts more tools at one's disposal, enabling more productive interaction with the environment, it also incurs a learning cost for each tool that quickly outweighs the advantage. For this reason, natural selection favors small behavioral repertoires that only include the most useful behaviors (section 4.4). At the same time, humans can use upwards of 500 voluntary muscles to home in on useful behavior patterns, despite the large cost of the search for such patterns (section 1.3.1). The goal of this section is to show that, through cultural evolution, humans are capable of performing a large number of potential behaviors yet employ only a small subset of useful ones. This may appear paradoxical—why be capable of many behaviors if only a few are used? Ultimately, different behaviors may be useful at different times, and cultural evolution can bring behaviors in and out of use much more quickly than genetic evolution. For example, our ancestors relied on the specialized motor skills involved in hunting and harvesting, but these have been largely replaced in modern society by other skills, such as typing and driving. The interplay between cultural evolution and human flexibility is further explored in chapter 16.

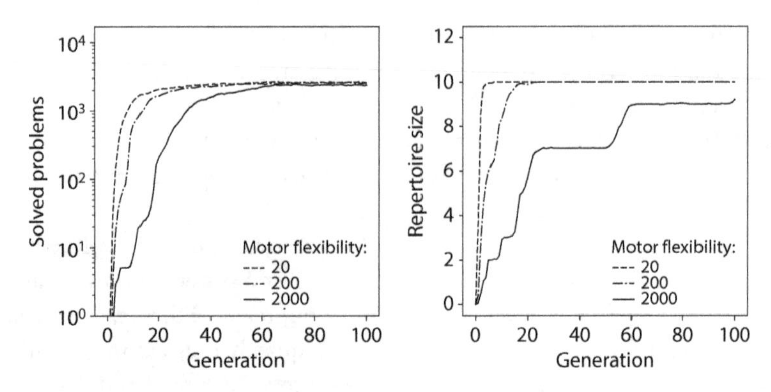

FIGURE 15.3. Relationship between cultural evolution and motor flexibility. Left: Result from three simulations in which 10 behaviors are required to solve problems in a model world, but individuals have potential repertoires of 20, 200, or 2000 behaviors. Even when most behaviors tried by individuals are useless, cultural evolution can select useful behaviors, leading to the ability to solve many problems. Right: Number of useful behaviors learned on average by individuals. Even with 2000 potential behaviors (99.5% useless), individuals can learn those that are useful to solve problems in the model world. Simulation setup is similar to figure 15.2; see section 1.5.5 for details.

To show that cultural evolution can identify useful skills within a large repertoire that contains many useless behaviors, we modify slightly the model world described in the previous section. We keep the number of useful problem-solving behaviors at 10 but allow each individual to use $m \geq 10$ behaviors. We assume that individuals are born without any behavioral skills, and that they acquire them either from others (social learning) or by drawing a randomly selected behavior from the potential repertoire (innovation). Simulation outcomes are shown in figure 15.3. The left panel shows that a population of individuals can learn to solve the same number of problems regardless of the number of potential behaviors, albeit more slowly when a vast fraction of the behaviors are useless. The right panel shows that when flexibility is higher than required, individuals can still learn the behaviors needed to solve all problems. Given the collective behavioral repertoire created by cultural evolution, we conclude that high behavioral flexibility is not incompatible with individual success within an environment.

15.4 Culture Can Support Mental Flexibility and Thinking

According to the model we proposed in chapter 8, animal intelligence derives mainly from a single, genetically determined learning algorithm, which we have referred to as chaining. This algorithm is efficient, economical, and ready to be deployed from birth. In contrast, we have argued since chapter 9 that human mental abilities are incredibly varied, and in large part learned over long periods of time. The puzzle of human mental flexibility is thus similar to that of behavioral flexibility: Without significant genetic guidance, how can humans discover useful mental processes among myriad flawed ones? (Just imagine trying to come up with calculus all on your own.) We believe that, similar to the acquisition of behavioral skills, individuals can acquire mental skills through a repertoire created by cultural evolution, thus reaping the advantages of mental flexibility but forgoing many of its costs. Indeed, the simulations in the previous section could describe mental as well as behavioral flexibility. Nodes in the world model can represent information states, such as what is known about a problem, and connections between nodes can represent mental operations that transform one information state to another. For example, the operation of addition would bring from the information state $5+3=?$ to the state $5+3=8$. This model is simplified, but it does suggest that, when mental abilities can be learned socially (as discussed in chapter 13), productive mental sequences can accumulate in a population and be transmitted to naive individuals.

Logical considerations and observations of human development provide important clues as to which mental skills are learned. Among the simulations in figure 15.2, the most successful one combines thinking with social learning. A fully functional thinking mechanism must contain many elements, including an appropriate representation of the environment, a sequential memory for reasoning and planning, strategies to evaluate evidence, and goals to strive for. Evidence suggests that social learning and cultural evolution make substantial contributions to most, if not all, elements of thinking (Gauvain 2001, Bjorklund and Causey 2017); and and we highlight some examples below, in addition to those presented in chapter 13.

15.4.1 Mental Representation

We have highlighted in section 15.2.2 that effective thinking relies on an appropriate representation of the environmental state, which can evolve culturally. For example, consider the representation of quantity. Many species demonstrate some ability to represent quantity and can discriminate arrays with different numbers of items (Brannon 2005), but only humans represent and manipulate quantity exactly, through the use of mathematical concepts. Pica et al. (2004) elucidated the cultural origin of this ability in experiments comparing French and Mundurukú speakers. The Mundurukú are an Amazonian indigenous people whose language only includes terms for 1–5, 10, "some" (roughly, between 5 and 10), and "many" (roughly, larger than 10). Only terms for 1–4 are used with precision. When asked to estimate magnitudes approximately, Mundurukú speakers performed as well as schooled French speakers, but they could not reliably solve addition and subtraction problems involving quantities larger than 4–5. (The problems were posed nonverbally in terms of arrays of dots.) Gordon (2004) and Frank et al. (2008) report similar results for Pirahã, another Amazonian people whose language has words only for 1/small size, 2/somewhat larger, and many. These findings indicate that exact arithmetic is a product of cultural evolution (Frank et al. 2008). Indeed, the evolution of number systems is apparent in historical times. For example, Roman and Mayan numerals are more cumbersome than modern place-based notation. More generally, developmental psychology emphasizes that children acquire many useful representations by interacting with adults, as seen in chapter 13.

Animals can also account for different properties of sensory information, depending on the context, but this ability is derived from a few fixed rules that cannot be altered. Namely, genetic predispositions influence learning about different properties of stimuli (section 4.2), and statistical learning can detect whether a specific feature is predictive of important outcomes and can thus alter its valence (Pearce 2008, Bouton 2016) (section 5.3). Likewise, learning of representation in animals depends on individuals' exposure to stimuli and does not have a culturally evolving component (section 5.2).

15.4.2 Mental Skills

There is ample evidence to suggest that advanced mental skills are products of cultural evolution. For example, computer scientists have formalized many problem-solving strategies in just the last few decades (Cormen et al. 2009), such as Dijkstra's (1959) algorithm to find the shortest path between two nodes in a graph (figure 15.1). Similarly, methods to optimize behavior in known and partially unknown environments have been devised over several decades (Bellman 1957, Bertsekas and Tsitsiklis 1996, Sutton and Barto 2018). But what about simpler planning and problem-solving strategies? Gauvain (2001) reviews evidence that adults guide children's learning of problem-solving skills by arranging tasks that are just beyond what the child can already do, as discussed in chapter 13. For example, Radziszewska and Rogoff (1988) showed that children who plan an errand with adults improve their planning skills more than children who plan with other children. Other examples can be found in Rogoff (1990) and Friedman and Scholnick (1997).

15.4.3 Adaptive Filtering

In the simulations above, we assumed that individuals would adopt and transmit to others only correct information about the world. Suppose, for example, that going from state i to state j requires performing behavior B_1, and that performing B_2 in i has no effect. Thus the sequences $i \to B_1 \to j$ and $i \to B_2 \to i \to B_2 \to i \to B_1 \to j$ accomplish the same effect. In our simulations, however, the second, less efficient sequence is automatically discarded. In reality, the utility of learning from others is seriously challenged by the danger of learning inefficient behavior and misleading information. Therefore, strategies to filter out inefficient solutions are vital to the long-term outcomes of cultural evolution (Enquist and Ghirlanda 2007). Cultural evolution theorists have often proposed that humans possess innate strategies of adaptive filtering, such as copying common behaviors or those displayed by successful individuals (Boyd and Richerson 1985, Henrich and Gil White 2001, Eriksson et al. 2007, Henrich 2015). Developmental psychology, however, provides evidence that adaptive filtering can itself evolve culturally. For example, children transition from believing others to evaluating evidence (Kuhn and Pearsall 2000), and become better at evaluating evidence with age (Amsel and Brock 1996). Furthermore, a child's ability to reason about evidence correlates with that of their parents (Luce et al. 2013).

15.4.4 Goals

Whereas animal goals are typically innate, such as building a nest or burying a nut (Ewer 1968), humans flexibly learn both short- and long-term goals, such as arriving to a meeting on time or saving for retirement. Anecdotally, we observe that most human activities are described in terms of goals. For example, we describe shopping in terms of what we need to buy, rather than in terms of the actions needed to shop. This focus may enable a more efficient understanding of the structure of tasks and the behavior of others (Loucks et al. 2017). In addition, we saw in section 12.2.2 that partitioning a complex task with subgoals can dramatically speed up learning, especially if appropriate

FIGURE 15.4. Seeking an intermediate goal can be productive even if the goal state is not rewarding in itself. In this example, pursuing an intermediate goal (circled) is a gateway to gain access to parts of the world that would otherwise be inaccessible.

goals can be acquired from knowledgeable individuals. There is a vast literature describing children's understanding and use of goals. Even six- to nine-month-olds appear capable of understanding the goals behind actions (Csibra and Gergely 2007, Woodward 2009), while children as young as three to four years can understand complex sequences of actions in which multiple goals are simultaneously pursued (Loucks et al. 2017). Therefore, there is ample opportunity for cultural evolution to act as a reservoir of useful goals in the same way as for useful behavior and mental procedures. For example, cultural evolution may result in specific goals that are not rewarding per se, but that enable individuals to access new opportunities (figure 15.4). Obtaining a hammer or a college degree, say, does not provide an immediate direct benefit, yet both are still useful goals, in appropriate environments.

15.5 Conclusion

In this chapter, we have used simple models to show that cultural evolution can accumulate useful solutions to problems, including solutions to information-processing problems that can be learned by individuals in the form of mental skills. The combination of thinking, social learning, and cultural evolution enables individuals and groups to discover orders of magnitude more productive sequences than if individuals were exploring the environment on their own through associative learning or thinking. This conclusion leads us to our last chapter, in which we address the difficult question of why full-fledged thinking and cultural evolution have become established only in humans.

16

Why Only Humans?

- In the human evolutionary transition, social transmission became the preponderant mode of information transmission.
- Language evolved as a new, more efficient channel for social transmission.
- Mental flexibility increased, making cumulative cultural evolution of mental skills possible.
- We suggest that these changes were primarily triggered by two unique circumstances:
  - An exceptionally long childhood, allowing extensive cost-free exploration.
  - An increase of useful sequential information in the social environment.
- These circumstances promoted the genetic evolution of sequential information processing abilities and mental flexibility.
- Gene-culture coevolution led to further changes to human childhood and inborn capacities, and to increased flexibility.
- Eventually, cultural evolution could proceed mainly on its own and shape its own evolutionary dynamics, creating many uniquely human mental skills.
- Several factors may have prevented other species from undergoing a similar evolutionary transition:
  - Associative learning is very efficient, and sufficient for many purposes.
  - Many species lack a protected juvenile period that is long enough to compensate for the costs of thinking.
  - The ecological niches occupied by most species may not require accurate sequential information processing.

16.1 Introduction

Let us summarize our arguments so far. Behavioral and mental mechanisms in animals are genetically controlled to a much greater extent than in humans. For example, a squirrel does learn, but this learning only perfects a set of standard behaviors that all squirrels learn (minor individual differences excepted). A human, while remaining human, can learn to be a singer, a surgeon, a shaman, and countless other things, with each behavioral skill underpinned by a large number of unique, learned mental skills. Tight genetic control of behavioral and mental mechanisms has both advantages and disadvantages: it reduces learning costs, but limits flexibility. As we have seen since our first chapter, learning behavioral sequences and processing sequential information incur particularly high costs, which oppose the evolution of these abilities. Humans, however, have evolved a new way of expanding flexibility and reducing learning costs: cultural evolution, including the cultural evolution of mental skills. In the long run, cultural evolution has resulted in more efficient behavior than is possible through strict genetic specification. Additionally, cultural evolution transforms itself, including transforming mechanisms for social transmission, mental world models, creativity, and adaptive filtering of social information.

Our hypothesis is that cultural evolution was made possible by a few genetically based improvements to human mental abilities, such as better representation and memory for sequential information and the ability to learn mental skills (chapter 2). However, once these changes were in place, cultural evolution proceeded largely autonomously from genetic evolution. In this final chapter, we first elaborate on our proposal that the evolution of human mental flexibility and human reliance on cultural information represent a major evolutionary transition. We then ask why humans are the only species that made this transition, analyzing potential hurdles to becoming a cultural species.

16.2 A Major Evolutionary Transition

In the influential book *The Major Transitions in Evolution*, Maynard Smith and Szathmáry (1995) suggested that the evolution of life on Earth has involved major events that changed not only the design of organisms but also the evolutionary process itself (see also Calcott and Sterelny 2011, Szathmáry 2015). Among the transitions they identified was the emergence of life itself, of chromosomes, DNA, the eukaryotic cell, and multicellularity. They also proposed that the evolution of human sociality and language was the latest transition. A key insight of Maynard Smith and Szathmáry is that all evolutionary transitions appear to share some common properties. For instance, smaller entities come together to form larger entities, such as in the evolution of prokaryotes into eukaryotic cells and in the evolution of multicellular life. Furthermore, the smaller entities often come to fulfill different roles and become unable to fully function on their own. Two examples are cell differentiation in multicellular organisms and division of labor in eusocial insects and human societies. Another property of evolutionary transitions is the emergence of new ways to organize and transmit information, such as nervous systems, cellular signaling systems, social learning, human language, and culture. Maynard Smith and Szathmáry also observed that the elements of a transition often continue to evolve significantly even after the transition, such as in the case of nervous systems (Kaas 2016). Finally, each major transition brings about new phenomena. For example, the evolution of sex brought about differences between the sexes, courtship, monogamy, and polygamy (Stearns 1987).

We agree with Maynard Smith and Szathmáry that human evolution is the embodiment of a major transition, in which the social transmission of information became a key evolutionary force. Indeed, many researchers have argued persuasively that genetic evolution is not sufficient to understand humans (Cavalli Sforza and Feldman 1981, Boyd and Richerson 1985, Tomasello 1999, Prinz 2012, Sterelny 2012, Heyes 2018). Henrich (2015, 314–317) effectively described this position by saying that we cannot understand humans as "really smart, though somewhat less hairy" chimpanzees: culture is as necessary to humans as water is to fish. The human evolutionary transition has led to a revolution that is both material and mental. Today, our lives would be unrecognizable without the material and nonmaterial culture we have created. The origin of this transition was the genetic evolution of mental abilities that made the invention and accumulation of culture possible. We have tentatively identified these abilities as innovations in sequential information processing and mental flexibility, and suggested that the ensuing cultural evolution eventually led to such traits as advanced social learning, thinking, cooperation, and language. In keeping with Maynard Smith and Szathmáry's framework, most elements of human uniqueness were not present at the beginning the transition, but self-organized as part of the transition itself. In the next two sections, we investigate the origins of the human evolutionary transition.

16.3 How Did It Start?

In considering why humans evolved unique mental abilities, an intriguing paradox arises. If we focus on any single environmental or biological factor in the life of ancestral hominids, such as ecological niche, social structure, or cognitive ability, we can typically find many species whose lives were in many ways similar to our ancestors', involving such activities as foraging, competition, and cooperation. This makes it difficult to understand why only one species underwent an evolutionary transition, or to identify a single causal factor. For example, according to the savanna hypothesis (Dart 1925), the drying up of forest habitat forced a change of lifestyle that triggered the evolution of human traits, such as upright walking and advanced tool making. However, such a broad ecological change would have similarly affected many species, and thus cannot be a complete explanation. Furthermore, if we are correct that the human evolutionary transition arose from domain-general abilities (chapter 2), it becomes even harder to pinpoint a single causal factor, as these abilities would have mattered in many domains, such as hunting, gathering, social learning, and cooperation. Despite these considerations, it remains valuable to discuss some environmental, social, and cultural factors that may have been important in human evolution.

Many researchers have theorized, in a similar fashion to the savanna hypothesis, that human evolution was driven primarily by a change in climate or habitat, which would have rendered existing behavior obsolete and triggered the evolution of new adaptations (Vrba 1995, Potts 1998, Godfrey-Smith 2001, Sterelny 2003, Ash and Gallup 2007, Tomasello 2014). Other researchers have focused on environmental variability, rather than on one specific change. Environmental variability decreases the effectiveness of genetic adaptation and may spur the evolution of faster ways to adapt, such as general intelligence (Godfrey-Smith 2001, Potts 2004) and culture (Boyd and Richerson 1985, Rogers 1988, Richerson and Boyd 2005, Maslin et al. 2015). Yet other hypotheses move beyond the physical environment, suggesting that changes in social organization, such as living in larger groups, stimulated the evolution of language and social intelligence (Humphrey 1976, Dunbar 1998, Sterelny 2003, Emery 2008). Some of these hypotheses focus on the challenges of group living, such as competition for resources (Humphrey 1976, Byrne and Whiten 1988, Povinelli and Preuss 1995), while others focus on its advantages, such as cooperation (Frith and Frith 2010, McNally and Jackson 2013). Finally, some researchers have seen a role for culture itself in setting human evolution apart from that of other primates. For example, control of fire and the subsequent availability of cooked food may have allowed more energy for brain growth (Robinson 1954, Jolly 1970, Aiello and Wheeler 1995, Wrangham 2009, Wrangham and Carmody 2010; but see Cornélio et al. 2016). This proposal is interesting in that it identifies a bottleneck (in this case, an energetic one) that may have prevented the evolution of intelligence in other animals, and may have been overcome by cultural evolution.

This brief survey of proposed factors in human evolution leads to the question of which environmental and social circumstances could have driven the genetic changes leading to human uniqueness. According to our hypothesis, it was an environment in which the ability to process sequential information and learn mental and behavioral skills paid off, without being hampered by combinatorial problems. Such an environment would have been complex in the sense of containing resources accessible only through long sequences of actions, but not necessarily in the sense of being unpredictable or variable over time. Indeed, in contrast to some of the hypotheses mentioned above, a stable environment might have offered more fertile ground for cultural evolution than a variable one, as even culture cannot cope with extreme environmental variability (Enquist and Ghirlanda 2007). The optimal environment appears to be a stable one, in which investing in learning and performing longer sequences can produce increasing returns. Using the terminology introduced in chapter 3, such an environment would have provided a favorable entry pattern for the evolution of sequential abilities and culture. The traditional objection to environmental stability as the cradle of culture is that optimal behavior in a stable environment can, in principle, be encoded genetically. However, culture can still have an edge in such an environment because it can discover productive behavior faster than genetic evolution, and also because it can carry more complex knowledge than could be encoded genetically.

The evolution of culture would also require a social environment in which valuable information could be obtained from others. In the next section, we consider how human evolution may have been spurred by increased opportunities for social learning.

16.4 Increased Learning Opportunities as a Path to Human Uniqueness

We do not deny that changes in environment and diet may have been important in human evolution, but we think that they provide only indirect cues to the evolution of uniquely human mental abilities. For example, it is hard to make specific predictions of how mental abilities would have changed given more energy for brain functioning. It would be valuable to establish more direct causal links between altered conditions and mental abilities, which we think is best achieved through formal and simulation models. Throughout this book, we have focused on the general problem of how animals find small islands of productivity in a sea of unproductive behavioral sequences (Tooby and Cosmides 1992). We have argued that the success of humans in

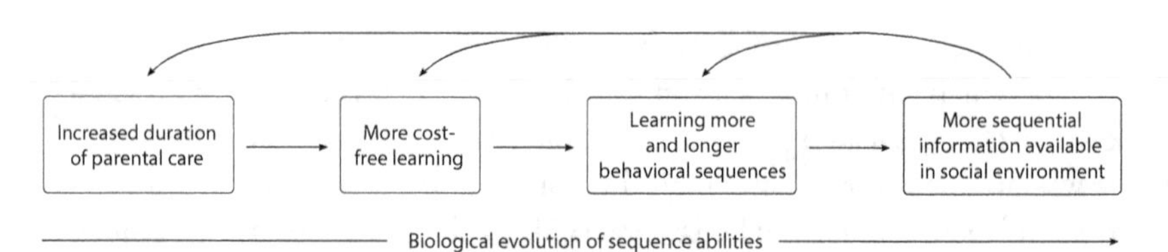

FIGURE 16.1. Our hypothesis for the initial steps of the human evolutionary transition. Increased duration of parental care increases opportunities for cost-free learning, leading to young individuals learning more and longer behavioral sequences. The latter creates a social environment even richer in information that the young can learn, creating a positive feedback with the preceding elements of the transition.

this difficult endeavor rests on the ability to learn mental skills that operate on sequential information and to transmit these skills culturally. Because of this focus on learning, it is natural for us to consider that a change in learning opportunities may have been instrumental in the evolution of human mental abilities. In particular, we focus on childhood learning, which can benefit from active tutoring and takes place in a protected environment that encourages exploration. Our hypothesis for the start of the human evolutionary transition is summarized in figure 16.1.

The typical mammalian life cycle can be conceptualized in three stages: infancy, in which the mother provides most or all nourishment; a juvenile stage, in which the young feeds on its own but is sexually immature; and a sexually mature adult stage. In humans, however, weaned children are still dependent on adults for food and protection, and young adults (adolescents) do not reproduce until 18–20 (in hunter-gatherer societies: Kaplan et al. 2000, Bogin and Smith 2012). The human life cycle thus provides greatly expanded opportunities for learning (Bruner 1972, Lewis 1983, Kaplan et al. 2000, Kaplan and Robson 2002, Bogin and Smith 2012, Sterelny 2012), which are necessary to acquire the behavioral, social, and mental skills necessary to function as adults (Bogin 2020, Rogoff 1990, Gopnik et al. 1999, Bjorklund and Causey 2017, Lew-Levy et al. 2017). During this time, humans display a dedication to teaching and learning that has no parallel in other species, including primates (Lancaster and Lancaster 1983, Bandura 1986, Herrmann et al. 2007, Sterelny 2012, Dean et al. 2012). Indeed, the very existence of active teaching in other apes is uncertain (Tomasello and Call 1997, Hoppitt et al. 2008). This greatly expanded active learning period results in longer reliance on adults, and yet this investment pays off in terms of higher productivity in adulthood (figure 16.2). Even in hunter-gatherers, this results

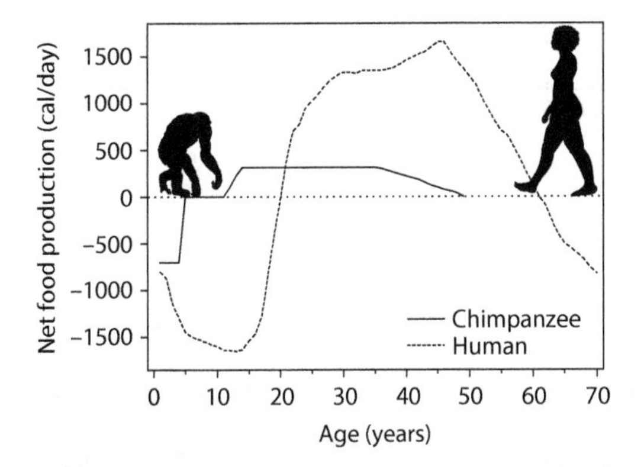

FIGURE 16.2. Differences in life span food production between chimpanzees at Gombe National Park and humans in existing traditional societies (Ache, Hiwi, and Hadza; redrawn using data from Kaplan and Robson 2002). Chimpanzees produce more calories than they consume by age seven, but humans do not become net producers until their early twenties. Human productivity is delayed by the time it takes to master sophisticated foraging techniques (Kaplan and Bock 2001, Kaplan et al. 2003), but eventually outpaces the productivity of chimpanzees.

in demographic growth beyond what other great apes can achieve (Kaplan et al. 2000, Bogin and Smith 2012).

In the following three subsections, we illustrate how a long childhood may be conducive to the evolution of uniquely human abilities, using the examples of sequential imitation and thinking.

16.4.1 Childhood Learning and Sequential Imitation

In section 15.2, we concluded that, when gathering knowledge about an environment, individual trial-and-error learning is much more limited than social learning and thinking (figure 15.2). However, the efficiency of trial-and-error learning increases with more learning opportunities. This effect is apparent in figure 16.3, in which we have used the same simulation settings as figure 15.2 to demonstrate how the number of learned sequences varies as a function of learning opportunities. The solid line represents an individual's ability to learn sequences. The dotted line represents the total number of sequences learned by a population of 100 individuals, given that each learns independently of the others. Because of chance differences in experience, each individual learns something unique, and the total knowledge in the population is much larger than the knowledge of each individual. This reservoir of knowledge, however, remains untapped unless individuals can learn from each other. The

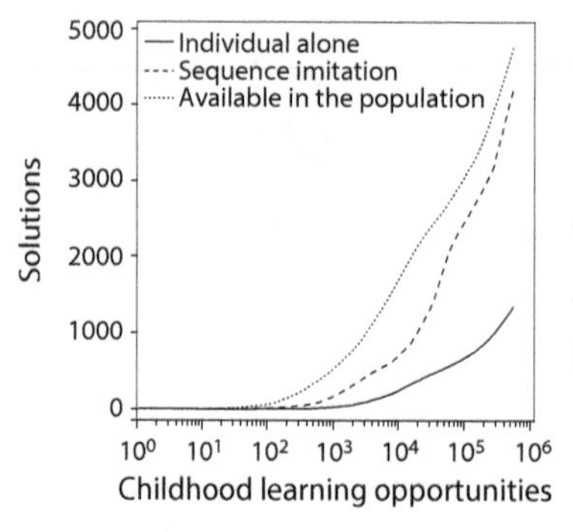

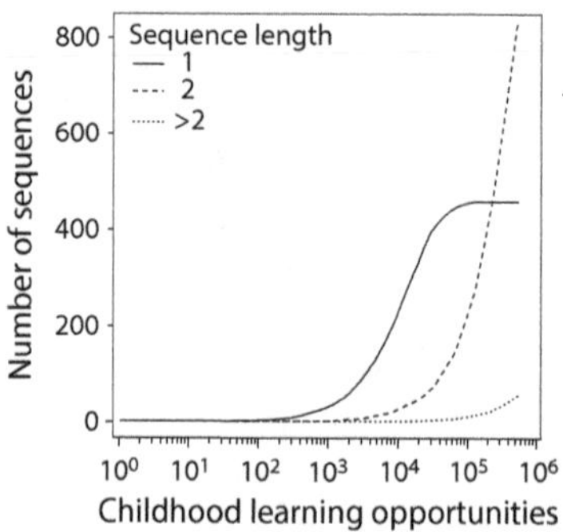

FIGURE 16.3. Childhood learning and sequential imitation. Left: Number of sequences learned as a function of childhood learning opportunities. Right: Number of sequences learned by an individual, according to sequence length. Simulation setup as in figure 15.2; see section 1.5.5 for details.

dashed line shows how much an individual could learn from others if capable of imitation. Note that there is no such individual in the simulated population; its addition is a thought experiment to investigate whether sequence imitation would be beneficial or not. When childhood is short, the benefit is slight; but with a longer childhood, the benefit can be substantial. Since sequence imitation requires sequential information processing (chapter 12), we can conclude that a longer childhood can promote the evolution of new mental abilities for sequential information processing by making imitation more beneficial. In fact, we can see in the right panel of figure 16.3 that with more learning opportunities the length of sequences learned by trial and error increases. Hence, an imitating individual needs sequence imitation abilities to be able to exploit the knowledge present in the population.

16.4.2 Childhood Learning and Thinking

In chapter 15, we showed that individuals capable of combining information from different experiences can learn much more about an environment than is possible through trial and error. Recombination, however, is not very effective when the individual has few learning opportunities, because there is little to be recombined. As discussed in chapter 11, learning a mental model of the world takes time. It is thus intuitive that the number of learning opportunities should strongly influence whether mental recombination of information is superior to trial-and-error learning. Using again the simulation setup from chapter 15 and figure 16.3, we show in figure 16.4 the dramatic effect of increased learning opportunities on the utility of recombination. The left panel shows that there is hardly any use of recombination until a sufficient number of learning opportunities is reached (roughly 1000 in our simulation), and

FIGURE 16.4. Childhood learning and mental models. Left: Percentage of the total number of sequences learned by an individual capable of recombining information (solid line), and percentage of connections between states learned (dashed line). Right: Average length of learned sequences. Simulation setup as in figure 15.2; see section 1.5.5 for simulation details.

also that past this number the usefulness of recombination increases sharply. (This is not a peculiarity of our world model; similar sharp transitions are incredibly common in network structures: Friedgut and Kalai 1996, Frieze and Karoński 2016.)

The right panel of figure 16.4 shows another interesting effect: when knowledge of the world is incomplete, thinking can still solve problems, but it uses unnecessarily long sequences. As an intuitive example, imagine navigating a city in which you know only half of the streets. You may still be able to get around, but you won't be using the best possible routes. With complete or nearly complete knowledge, however, the shortest sequences can be found. Thus, while even approximate mental models can be useful, there is an enduring incentive to improve mental models.

We conclude from these results that thinking is unlikely to be of much use to species that lack a prolonged learning period, because of the need to learn a mental model that can be used for thinking. The bottleneck is even more severe if we consider that thinking requires sophisticated sequence-processing abilities in order to mentally recombine sequences of actions.

16.4.3 Childhood Learning, Language, and Symbolic Codes

Language is one of the most striking human abilities and the most common symbolic code used by humans. We have not devoted a specific chapter to language, but we have included it in our discussion from chapter 9 onward. In particular, chapter 12 covers how symbolic codes can boost social learning, while chapter 13 discusses language as one of the fundamental, uniquely human mental abilities acquired during childhood.

Our hypothesis of human evolution considers language as mainly a product of cultural evolution, which has emerged in cultural coevolution with other uniquely human mental abilities, such as advanced imitation abilities and thinking. In fact, our discussion of codevelopment in section 13.3.3 suggests that language could not have evolved without coevolving with these other mental abilities (similarly, Deacon [1997] suggests that language depends on symbolic thought). Indeed, it is tempting to speculate that language rests on the same genetic foundations as the other uniquely human mental abilities, namely, on improved representation, memory of sequential information, and the ability to learn mental skills. We do not exclude the fact that specific genetic adaptations for language exist (even beyond the vocal apparatus), but we think it likely that these have evolved after the first language abilities emerged. One reason for this belief is that simple, yet effective, communication seems possible based on general-purpose sequential abilities (Christiansen and MacDonald 2009, Heyes 2018).

Considering the coevolution of language and childhood, as we did above for imitation and thinking, also hints that the early stages of language evolution may have been primarily cultural. Even in its early stages of evolution, language would have conferred at least two fitness advantages. One is improved transmission of information from parent to offspring, and the other is improved coordination among group members. These advantages would have increased as the social environment became richer in information, due to the coevolution of childhood, imitation, and thinking. At some point, the social environment would have contained a significant number of behavioral sequences, such that a symbolic system would have been beneficial for their transmission. Thus one of the early steps toward language evolution may have been the cultural evolution of symbolic codes to accomplish deferred imitation of behavioral sequences, which would have been possible following the genetic evolution of an adequate sequential memory.

Finally, the idea that language is primarily a cultural product can explain one of language's most puzzling features: no other species appears to be close to developing a full-fledged language capacity (Deacon 1997). While it is partly a matter of definition whether one considers sophisticated animal communication systems as a "language" (such as in bees or prairie dogs: Von Frisch 1967, Slobodchikoff 2002), there are definitely no animals with the linguistic ability of even a two- or three-year-old child. If language is primarily a cultural innovation, however, it could have arisen practically instantaneously, on the timescale of genetic evolution. In other words, human language left animal communication systems behind once it started to evolve culturally rather than genetically. Indeed, a similar argument can be made for all uniquely human mental abilities, supporting the idea that they are primarily the product of cultural evolution (chapter 13).

16.5 Hurdles to the Human Evolutionary Transition

The initial stages of human evolution outlined in the preceding section may also be observed in some animals, particularly in the long learning periods and large cultural repertoires of the great apes. However, only humans have fully transitioned to the use of mental flexibility and cultural evolution as the primary mode of adaptation. In this section, we consider three hurdles that may have prevented other species from undergoing a similar transition: combinatorial learning costs, the amount of information available socially, and capacities for adaptive filtering.

The first hurdle concerns one of our main themes: advanced behavioral and mental skills require long sequences of actions and mental operations, which are hard to learn unassisted (chapter 1). In the same way that the cost of behavioral flexibility may outweigh its benefits (chapter 4), mental flexibility may not pay off if there is no guidance about which mental operations are useful (chapter 15). We have concluded that, due to combinatorial learning costs, the evolution of mental flexibility and culture are inseparable. Mental flexibility requires culture to provide information about which mental skills are actually useful, and culture requires mental flexibility for the development and transmission of such skills. Thus most animals are prevented from becoming fully cultural simply because they do not have enough time to invest in learning. For example, a mentally flexible animal with a short or solitary life would have limited possibilities to learn mental skills. The picture of animal intelligence that emerges from chapter 8 is not one of mental flexibility, but one of a preprogrammed learning algorithm—chaining—that is supported by genetic predispositions as well as specialized information sources and memory mechanisms (chapter 5).

The second hurdle concerns the amount of social information that can be learned by individuals. We have seen in chapter 6 that guided associative learning is sufficient for social learning of simple behavior. Therefore, there may not be strong incentive to depart from associative learning unless the social environment includes complex, productive behavior. In other words, advanced mental abilities for learning socially are not of much use when there is little social information (Boyd and Richerson 1996, Enquist et al. 2008, Muthukrishna et al. 2018). For example, figure 16.3 shows that the utility of faithful imitation may be very limited in the initial stages of its evolution. Figure 16.4 makes a similar point about thinking abilities.

The third hurdle concerns the necessity of adaptive filtering, that is, the preferential retention of productive information (chapter 15). In associative learning, adaptive filtering occurs because rewarded behavior increases in frequency while unrewarded behavior is eventually abandoned. This is possible for simple behaviors with direct consequences, but not for long behavioral sequences, such as farming or building a permanent dwelling, whose success or failure depends on a multitude of factors. Indeed, because of limitations in sequential memory, animals may have difficulties evaluating the adaptiveness of even relatively short behavioral sequences, such as multistep foraging techniques (chapters 5 and 8). Unless unproductive behavior can be filtered out, there may be little advantage to becoming a mentally flexible, cultural species (Enquist and Ghirlanda 2007).

To appreciate the significance of these hurdles, consider the great apes. Chimpanzees, orangutans, and gorillas have the largest known animal cultures, with about 15–30 cultural behaviors documented in each local population and a few dozen in each species (Whiten et al. 1999, van Schaik et al. 2003, Wich et al. 2010, Robbins et al. 2016). At first sight, it appears that these animals would benefit from increased mental flexibility, for example, to learn more effective teaching or imitation or to improve on what they learn from others. The three hurdles just discussed, however, suggest otherwise. First, learning the relatively simple behaviors already takes a large part of a great ape's juvenile period (section 3.8.4). If increasing flexibility also increases learning times, the apes might end up being able to learn less rather than more. Second, known cultural behaviors among great apes appear simple enough that they can be rediscovered by a single individual (Neadle et al. 2017, Bandini and Tennie 2017, Bandini and Harrison 2020, Acerbi et al. 2020). This suggests that there may be only a weak selection pressure to invest more in mental abilities for social learning or for improving what is learned (Gabora 1998, Gabora and Steel 2017). Finally, the great apes may lack effective adaptive filtering of complex behavior. For example, Boesch (1995) observed that some chimpanzee cultural behaviors provide no obvious benefit. In the case of leaf clipping, which is the act of ripping pieces of leaves with the teeth, the only known effect is that it makes a distinctive sound (the leaves are not eaten). Without effective filtering, unproductive cultural behavior can accumulate and lower fitness. These considerations suggest that the selection pressure for increasing mental flexibility in the great apes may be weak.

16.6 Gene-Culture Coevolution of Human Uniqueness

Once uniquely human mental capacities emerged through genetic evolution (chapter 10) and culture evolved beyond what occurs in animals (chapter 8), it is likely that genes and culture coevolved, that is, influenced and reinforced each other's evolution (Lumsden and Wilson 1981, Boyd and Richerson 1985, Durham 1991, Henrich and McElreath 2007, Laland 2008). We suggest that gene-culture coevolution caused genetic changes mainly in three areas related to inborn mental capacities, as discussed next. (We do not discuss morphological and physiological changes in other areas, such as the coevolution of dairy products and lactose tolerance; see Bersaglieri et al. 2004, Itan et al. 2009.)

16.6.1 Childhood

It is likely that gene-culture coevolution further increased the duration of childhood and the motivation for social learning in children and for tutoring in parents, beyond what genetic evolution alone had already produced (Kaplan et al. 2000, Konner et al. 2010). The selection pressure for these further increases would be the benefits of learning from others during childhood, resulting in more productive adult behavior that would promote the fitness of both the individual and their parents. Note, however, that both genetic and cultural forces can affect these areas, and so it may be difficult to separate their effects. For example, the time devoted to learning as well as the time spent with parents have increased considerably in the last few hundred years, purely for cultural reasons.

16.6.2 Inborn Mental Skills

Inborn mental skills that may have been amplified by gene-culture coevolution include domain-general support for sequential information processing as well as mental flexibility. While we have hypothesized that language and thinking are mainly the product of domain-general mechanisms, domain-specific skills may also have evolved in this phase, for example, to further support thinking and language, such as to improve sound production (chapters 10 and 13). At the same time, we do not believe that gene-culture coevolution introduced fundamental, domain-specific innovations to the human mental architecture, primarily because domain specificity would decrease mental flexibility and thus would hamper the cultural evolution of mental skills.

16.6.3 Increased Flexibility

Perhaps the most important inborn change caused by gene-culture coevolution might be a significant increase in mental flexibility, resulting from the removal of inborn predispositions that were once necessary to guide learning (Heyes 2018). In this phase of human evolution, we would thus see more combinatorial problems being solved through cultural information (chapter 10) rather than through genetic predispositions (chapter 4). Indeed, without a dramatic increase in mental flexibility, the scope for human cultural evolution would be severely limited (Richerson and Boyd 2005, Mesoudi 2011, Henrich 2015). Concretely, increased mental flexibility would have allowed connecting stimuli with behavioral responses more freely, acquiring new mental skills, and also learning what goals one should strive for. In general, human behavior is today less controlled by genetics in many domains, compared to the behavior of other animals. For example, although there are genetic predispositions in our initial reactions to food, such as when we taste sweet and bitter, we can learn to like almost any taste. Further, there are effectively no remaining predispositions in how to obtain food. Human foraging varies culturally in many ways, including activities like hunting and gathering, growing one's own food, industrial farming and breeding, cooking and preserving food by myriad different methods, shopping in stores, and visiting restaurants.

16.7 The Rise of Cultural Evolution

Once inborn cultural capacities and sufficient mental flexibility were established, cultural evolution of mental and behavioral skills could proceed at a much faster pace, and with more or less open-ended potential (Gabora 1997, 1998, Gabora and Steel 2017, 2020). As cultural evolution started, a new evolutionary process had seen the light of day, and, as genetic hurdles were overcome, mental and bodily constraints were compensated for by inventing tools, machines, and computers. The rapid pace of cultural evolution explains how a diversity of cultural phenomena has emerged on Earth rather recently, such as reading and writing, mathematics, agriculture, pottery and many other crafts, technologies such as metallurgy and computers, and science. These phenomena cannot be attributed to genetic evolution.

How long have the inborn capacities for culture been in place? To make this question more explicit, let us consider a particular skill, say, literacy. Barring pathologies, all human children can learn to read and write, even children whose parents and ancestors never possessed these mental skills. Written symbols are at least 5000 to 6000 years old (Daniels and Bright 1996, Robinson 1999), so we know at least some humans could learn to read and write back then. We can go further back in time, however, and ask whether a child from 25,000 years ago would also be able to learn to read and write. And what about a child that lived 150,000 or 300,000 years ago? The latter ages are relevant because genetic studies indicate that *Homo sapiens sapiens* diverged early into several subgroups that since have had no or only limited contact with each other (figure 16.5; Hublin et al. 2017, Schlebusch et al. 2017). If we assume a single origin of human cultural capacities, these capacities would

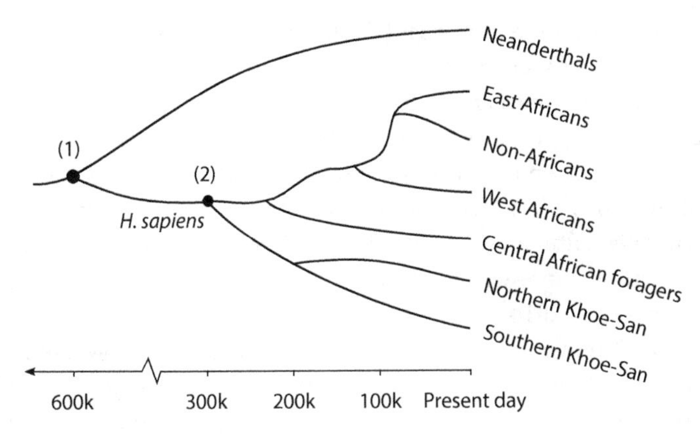

FIGURE 16.5. A simple phylogenetic tree of human evolution. (1) The major genetic split between the Neanderthal and *sapiens* lineages is estimated at around 600k BP (years before present). (2) Genetic divergence between *sapiens* groups that still exist today are estimated to have occurred between 140k and 300k BP. Phylogeny and dates from Schlebusch et al. (2017).

have been fully developed a long time ago, that is, they would be present at least in the most recent common ancestor of all *H. s. sapiens* (Henshilwood and Marean 2003, Lind et al. 2013). Indeed, it is thought provoking that these inborn capacities could have been fully developed even earlier, as suggested by the increasing number of similarities found between *H. sapiens* and *H. neanderthalensis* from both archaeological (Hayden 1993, Speth 2004, Zilhão et al. 2006, Langley et al. 2008, Morin and Laroulandie 2012, Hoffmann et al. 2018) and genetic studies (Krause et al. 2007, Green et al. 2010, Yotova et al. 2011, Prüfer et al. 2014).

A different idea is that human cultural capacities are much more recent, perhaps as recent as 40,000 years, and that they spread across human populations later through migration and mating (Tattersall 1995, Ambrose 1998, Wolpoff et al. 2000, Klein 2001, Bar Yosef 2002, Klein 2008). Today, this idea is commonly rejected (Lind et al. 2013), for example, based on observations of art and advanced stone tools as old as 200,000 years (McBrearty and Brooks 2000, Henshilwood et al. 2002, 2009, Nowell 2010, Aubert et al. 2014, Hoffmann et al. 2018).

If humans did possess the necessary mental capacities for learning to read and write as early as 300,000 years ago, how can we explain that reading and writing only emerged some 6000 years ago? Similar questions can be asked about almost all uniquely human phenomena, which has been referred to as the "sapient paradox" (Renfrew 1996, 2008). However, if these phenomena depend on cumulative cultural evolution, this temporal pattern is expected.

When changes occur to fundamental aspects of the evolutionary process, the most dramatic consequences may materialize only long after (chapters 10 and 15; Maynard Smith and Szathmáry 1995). For example, DNA initially contained only a few genes and was responsible for very simple life-forms. Similarly, at the dawn of human culture, there were only a few rudimentary cultural elements. The process of cultural evolution itself was also inefficient, but eventually cultural evolution self-organized into a much more efficient system to create, represent, transmit, and use cultural information (Sterelny 2012, Acerbi et al. 2014).

The transition between an initial phase of exceedingly slow cultural evolution and a later phase of rapid growth can be understood from cumulative processes in which the creation of new culture depends on already existing culture (Ogburn 1950, Tomasello 1999, Laland et al. 2001, Odling Smee et al. 2003, Boyd and Richerson 2005, Heyes 2009, Enquist et al. 2008, 2011, Sterelny 2012). A number of mathematical models have identified thresholds in this transition, below which culture would have existed in a relatively static state, similar to culture in great apes. For example, a small population, or even a large but fragmented population, is likely unable to generate sustained cultural growth (Boserup 1965, Henrich 2004, Powell et al. 2009, Strimling et al. 2009, Kobayashi and Aoki 2012, Fogarty et al. 2013, Lind et al. 2013, Kolodny et al. 2015, Shennan 2015, Creanza et al. 2017). Similarly, Gabora (1997, 1998) has identified cognitive thresholds that are very much in line with our perspective (see also Gabora and Steel 2017, 2020). In these models, individuals initially have very small mental models, as these are learned from a very simple social and cultural environment. Such small mental models would result in a limited ability to generate new mental and behavioral skills, similar to our simulations in section 16.4. As new mental elements are painstakingly added, however, a threshold is eventually reached, at which point mental models and mental skills are sophisticated enough to enable much faster generation of mental and behavioral innovations.

16.8 Genes Still Matter

Our hypotheses of human evolution do not imply that genes have no impact on human mental and behavioral skills. In fact, we share most aspects of brain and mental mechanisms with many other species. In particular, associative learning is still important in humans, as many of our decisions are fast and not based on deliberate thinking (Kahneman 2011). The role of associative learning is particularly pronounced in certain domains in which genetic predispositions are still important, for example, in learning about stimuli that elicit fear, anxiety, and avoidance (DeLoache and LoBue 2009, LoBue and Rakison 2013). Understanding how genetic predispositions and associative learning still influence human behavior is crucial in clinical psychology (Bernstein 1999, Haselgrove and Hogarth 2013).

More relevant to our arguments is that uniquely human mental skills would not exist unless some genetic differences existed between humans and animals. Where our views of human evolution differ from others is in the effects of these differences, which we have tentatively identified in increased mental flexibility and domain-general mechanisms for sequential information processing, rather than in producing specific mental skills (chapters 2 and 10). The latter, we argued, are the consequence of learning and cultural evolution. Thus we reach similar conclusions as Jablonka and Lamb (2005), Prinz (2012), and Heyes (2018), and propose that human universals derive from learning in similar cultural and physical environments, rather than from genetically determined mental skills (Brown 1991, Pinker 2002).

While we think there is, typically, little inborn control of how humans solve problems and fulfill goals, we also note that much human behavior is still driven in large part by inborn motivational and emotional systems that we share with other animals, and that cause us to strive for water, food, mates, safety, and so on. While goals are also subject to cultural evolution (such as "serve your country" or "be at the top of your college class"), inborn goals continue to affect human life, including culture.

16.9 Conclusion: A Mental Revolution

We remarked in chapter 9 that mental worlds created by humans are in every respect as impressive as our material culture. In successive chapters, we have traced the origin of these mental worlds to unique sequential learning abilities and mental flexibility. Through cultural evolution, these abilities have resulted in the creation of new mental skills and the gathering of information embedded in mental worlds. Indeed, human material culture is itself, to a large extent, a consequence of learned mental skills.

Our first major conclusion in this book is that mental flexibility and culture are inextricably linked. Mental flexibility would not be beneficial without culture providing a reservoir of useful mental skills, and culture could not exist without the mental flexibility necessary to create and learn such skills. We believe that the cultural evolution of mental skills is a more viable explanation for human uniqueness than assuming either a myriad of largely innate mental modules or a few major innate mechanisms for general intelligence. The idea of human abilities as a Swiss army knife (Tooby and Cosmides 1992) with many innate tools does not explain the emergence of new mental and behavioral phenomena, while the idea of general intelligence does not explain how a few fixed mental mechanisms can generate the staggering diversity of human mental skills (chapter 10).

Our second major conclusion is that animal and human intelligence represent alternative solutions to combinatorial dilemmas in learning and decision making. They are both powerful, but only human intelligence relies on domain-general thinking, that is, on recombination of sequential information in all aspects of human life. Animals rely instead on associative learning aided by genetic information and specialized memory systems. The transition from guided associative learning to thinking is a difficult one to make, because associative learning is very efficient, and because thinking itself carries costs that cannot be offset without extensive childhood learning and access to cultural information. We have not presented a complete theory of how this evolutionary transition occurred, yet we hope our focus on sequential information processing and cost-benefit analyses adds a useful perspective to the classic question of what sets us apart from other animals.

REFERENCES

Acerbi A, Ghirlanda S, Enquist M. 2014. Regulatory traits: Cultural influences on cultural evolution. In Cagnoni S, Mirolli M, Villani M, eds., *Evolution, complexity and artificial life*, 135–147. Springer: Berlin.

Acerbi A, Mesoudi A. 2015. If we are all cultural Darwinians what's the fuss about? Clarifying recent disagreements in the field of cultural evolution. *Biology & Philosophy* 30(4): 481–503.

Acerbi A, Snyder WD, Tennie C. 2020. Ape cultures do not require behavior copying. *bioRxiv*. https://doi.org/10.1101/2020.03.25.008177.

Adams CD. 1980. Post-conditioning devaluation of an instrumental reinforcer has no effect on extinction performance. *Quarterly Journal of Experimental Psychology* 32(3):447–458.

Adams CD, Dickinson A. 1981. Instrumental responding following reinforcer devaluation. *Quarterly Journal of Experimental Psychology* 33(2):109–121.

Addessi E, Crescimbene L, Visalberghi E. 2008. Food and token quantity discrimination in capuchin monkeys (*Cebus apella*). *Animal Cognition* 11(2):275–282.

Aiello LC, Wheeler P. 1995. The expensive-tissue hypothesis: The brain and the digestive system in human and primate evolution. *Current Anthropology* 36(2): 199–221.

Aisner R, Terkel J. 1992. Ontogeny of pine cone opening behaviour in the black rat, *Rattus rattus*. *Animal Behaviour* 44:327–336.

Akins CK, Klein ED, Zentall TR. 2002. Imitative learning in Japanese quail (*Coturnix japonica*) using the bidirectional control procedure. *Animal Learning & Behavior* 30(3):275–281.

Akins CK, Zentall TR. 1996. Imitative learning in male Japanese quail (*Coturnix japonica*) using the two-action method. *Journal of Comparative Psychology* 110(3):316–320.

Allen TA, Fortin NJ. 2013. The evolution of episodic memory. *Proceedings of the National Academy of Sciences* 110(2):10379–10386.

Allison ML, Reed R, Michels E, Boogert NJ. 2020. The drivers and functions of rock juggling in otters. *Royal Society Open Science* 7(5):200141.

Altmann SA. 1974. Baboons, space, time, and energy. *American Zoologist* 14(1):221–248.

Ambrose SH. 1998. Chronology of the later Stone Age and food production in East Africa. *Journal of Archaeological Science* 25(4):377–392.

Amsel E, Brock S. 1996. The development of evidence evaluation skills. *Cognitive Development* 11(4):523–550.

Amundson JC, Miller RR. 2008. CS-US temporal relations in blocking. *Learning & Behavior* 36(2):92–103.

Anderson JR. 1982. Acquisition of cognitive skill. *Psychological Review* 89(4):369.

Anderson JR. 1993. *Rules of the mind*. Mahwah, NJ: Lawrence Erlbaum.

Ando J, Ono Y, Wright MJ. 2001. Genetic structure of spatial and verbal working memory. *Behavior Genetics* 31(6):615–624.

Angermeier WF. 1960. Some basic aspects of social reinforcements in albino rats. *Journal of Comparative and Physiological Psychology* 53(4):364.

Anthony W. 1959. The Tolman and Honzik insight situation. *British Journal of Psychology* 50(2):117–124.

Aoki K, Wakano JY, Feldman MW. 2005. The emergence of social learning in a temporally changing environment: A theoretical model. *Current Anthropology* 46:334–340.

Aplin LM. 2019. Culture and cultural evolution in birds: A review of the evidence. *Animal Behaviour* 147:179–187.

Aplin LM, Farine DR, Morand-Ferron J, Cockburn A, Thornton A, Sheldon BC. 2015. Experimentally induced innovations lead to persistent culture via conformity in wild birds. *Nature* 518(7540):538–541.

Arbib MA, ed. 2003. *The handbook of brain theory and neural networks*. 2nd ed. Cambridge, MA: MIT Press.

Ash J, Gallup GG. 2007. Paleoclimatic variation and brain expansion during human evolution. *Human Nature* 18(2):109–124.

Ashcraft MH, Fierman BA. 1982. Mental addition in third, fourth, and sixth graders. *Journal of Experimental Child Psychology* 33(2):216–234.

Atance CM, Meltzoff AN. 2005. My future self: Young children's ability to anticipate and explain future states. *Cognitive Development* 20(3):341–361.

Aubert M, Brumm A, Ramli M, Sutikna T, Saptomo EW, Hakim B, Morwood MJ, van den Bergh GD, Kinsley L, Dosseto A. 2014. Pleistocene cave art from Sulawesi, Indonesia. *Nature* 514(7521):223–227.

Axelrod R. 1984. *The evolution of cooperation*. New York: Basic Books.

Axelrod R, Hamilton W. 1981. The evolution of cooperation. *Science* 211:1390–1396.

Babb SJ, Crystal JD. 2006. Episodic-like memory in the rat. *Current Biology* 16(13):1317–1321.

Balda RP, Kamil AC. 1992. Long-term spatial memory in Clark's nutcracker, *Nucifraga columbiana*. *Animal Behaviour* 44(4):761–769.

Balda RP, Pepperberg IM, Kamil AC, eds. 1998. *Animal cognition in nature*. New York: Academic Press.

Balleine BW. 1992. Instrumental performance following a shift in primary motivation depends on incentive learning. *Journal of Experimental Psychology: Animal Behavior Processes* 18(3):236–250.

Balleine BW, Dickinson A. 1998. Goal-directed instrumental action: Contingency and incentive learning and their cortical substrates. *Neuropharmacology* 37(4):407–419.

Balleine BW, Dickinson A. 2005. Effects of outcome devaluation on the performance of a heterogeneous instrumental chain. *International Journal of Comparative Psychology* 18(4): 257–272.

Bandini E, Harrison RA. 2020. Innovation in chimpanzees. *Biological Reviews* 95(5):1167–1197.

Bandini E, Tennie C. 2017. Spontaneous reoccurrence of "scooping," a wild tool-use behaviour, in naïve chimpanzees. *PeerJ* 5:e3814.

Bandura A. 1986. *Social foundations of thought & action: A social cognitive theory*. Eaglewood Cliffs, NJ: Prentice Hall.

Bar Yosef O. 2002. The Upper Paleolithic revolution. *Annual Review of Anthropology* 31:363–393.

Barker KB, Povinelli DJ. 2019. Anthropomorphomania and the rise of the animal mind: A conversation. *Journal of Folklore Research* 56(2–3):71–90.

Barnosky AD, Matzke N, Tomiya S, Wogan GO, Swartz B, Quental TB, Marshall C, McGuire JL, Lindsey EL, Maguire KC, et al. 2011. Has the Earth's sixth mass extinction already arrived? *Nature* 471(7336):51.

Barrett HC. 2015. Modularity. In Zeigler-Hill V, Welling LLM, Shackelford T, eds., *Evolutionary perspectives on social psychology*, 39–49. Cham, Switzerland: Springer.

Barry RA, Graf Estes K, Rivera SM. 2015. Domain general learning: Infants use social and non-social cues when learning object statistics. *Frontiers in Psychology* 6:551.

Bauer PJ, Wenner JA, Dropik PL, Wewerka SS, Howe ML. 2000. Parameters of remembering and forgetting in the transition from infancy to early childhood. *Monographs of the Society for Research in Child Development*, i–213.

Baum EB. 2004. *What is thought?* Cambridge, MA: MIT Press.

Baum WM. 1974. On two types of deviation from the matching law: Bias and undermatching 1. *Journal of the Experimental Analysis of Behavior* 22(1):231–242.

Becker GS, Murphy KM. 1992. The division of labor, coordination costs, and knowledge. *Quarterly Journal of Economics* 107(4):1137–1160.

Bednekoff PA, Balda RP, Kamil AC, Hile AG. 1997. Long-term spatial memory in four seed-caching corvid species. *Animal Behaviour* 53(2):335–341.

Bekoff M. 2008. *The emotional lives of animals: A leading scientist explores animal joy, sorrow, and empathy–and why they matter*. Novato, CA: New World Library.

Bekoff M, Allen C, Burghardt G, eds. 2002. *The cognitive animal*. Cambridge, MA: MIT Press.

Bellman RE. 1957. *Dynamic progamming*. Princeton NJ: Princeton University Press.

Bellman RE. 1961. *Adaptive control processes: A guided tour*. Princeton NJ: Princeton University Press.

Benjamin A, Shermer M. 2006. *Secrets of mental math: The mathemagician's guide to lightning calculation and amazing math tricks*. New York: Three Rivers Press.

Bennett AT. 1996. Do animals have cognitive maps? *Journal of Experimental Biology* 199(1):219–224.

Benson JB. 1997. The development of planning: It's about time. In Friedman S, Scholnick EK, eds., *The developmental psychology of planning: Why, how, and when do we plan?*, 43–75. New York: Psychology Press.

Bentley-Condit VK, Smith E. 2010. Animal tool use: Current definitions and an updated comprehensive catalog. *Behaviour* 147(2):185–221.

Berger PL, Luckman T. 1967. *The social construction of reality*. New York: Anchor Books.

Bering JM. 2004. A critical review of the "enculturation hypothesis": The effects of human rearing on great ape social cognition. *Animal Cognition* 7(4):201–212.

Beritashvili IS. 1972. Phylogeny of memory development in vertebrates. In Karczmar A, Eccles JC, eds., *Brain and Human Behavior*, 341–349. Berlin: Springer-Verlag.

Berk LE. 2013. *Child development*. 9th ed. Boston: Pearson.

Berlyne D, Slater J. 1957. Perceptual curiosity, exploratory behavior, and maze learning. *Journal of Comparative and Physiological Psychology* 50(3):228.

Berna F, Goldberg P, Horwitz LK, Brink J, Holt S, Bamford M, Chazan M. 2012. Microstratigraphic evidence of in situ fire in the Acheulean strata of Wonderwerk Cave, Northern Cape province, South Africa. *Proceedings of the National Academy of Sciences* 109(20):E1215–E1220.

Bernstein IL. 1999. Taste aversion learning: A contemporary perspective. *Nutrition* 15(3): 229–234.

Bersaglieri T, Sabeti PC, Patterson N, Vanderploeg T, Schaffner SF, Drake JA, Rhodes M, Reich DE, Hirschhorn JN. 2004. Genetic signatures of strong recent positive selection at the lactase gene. *American Journal of Human Genetics* 74(6):1111–1120.

Bertsekas DP. 2012. *Dynamic programming and optimal control*, vol. 1. Cambridge, MA: Athena Scientific.

Bertsekas DP, Tsitsiklis JN. 1996. *Neuro-dynamic programming*. Cambridge, MA: Athena Scientific.

Biebach H, Gordijn M, Krebs JR. 1989. Time-and-place learning by garden warblers, *Sylvia borin*. *Animal Behaviour* 37:353–360.

Bird CD, Emery NJ. 2009. Insightful problem solving and creative tool modification by captive nontool-using rooks. *Proceedings of the National Academy of Sciences* 106(25):10370–10375.

Bird CD, Emery NJ. 2010. Rooks perceive support relations similar to six-month-old babies. *Proceedings of the Royal Society B* 277(1678):147–151.

Bischof HJ. 1994. Sexual imprinting as a two-stage process. In Hogan JA, Bolhuis JJ, eds., *Causal mechanisms of behavioural development*, 82–97. Cambridge, UK: Cambridge University Press.

Bitterman ME, LoLordo VM, Overmier JB, Rashotte ME, eds. 1979. *Animal learning: Survey and analysis*. New York: Plenum Press.

Bjorklund DF, Causey KB. 2017. *Children's thinking: Cognitive development and individual differences*. Thousand Oaks, CA: Sage.

Blair-West J, Coghlan J, Denton D, Nelson J, Orchard E, Scoggins B, Wright R, Myers K, Junqueira C. 1968. Physiological, morphological and behavioural adaptation to a sodium deficient environment by wild native Australian and introduced species of animals. *Nature* 217(5132):922.

Blaisdell AP, Sawa K, Leising KJ, Waldmann MR. 2006. Causal reasoning in rats. *Science* 311(5763):1020–1022.

Blodgett HC. 1929. The effect of the introduction of reward upon the maze performance of rats. *University of California Publications in Psychology* 4(8):113–134.

Bloom JM, Capaldi E. 1961. The behavior of rats in relation to complex patterns of partial reinforcement. *Journal of Comparative and Physiological Psychology* 54(3): 261–265.

Bloomfield TM. 1972. Reinforcement schedules: Contingency or contiguity? In Gilbert RM, Millenson JR, eds., *Reinforcement: Behavioral analyses*, 165–208. New York: Academic Press.

Blough DS. 1990. Form similarity and categorization in pigeon visual search. In Commons ML, Herrnstein RJ, Kosslyn SM, Mumford DB, eds., *Quantitative analyses of behavior*, 129–143. New York: Psychology Press.

Boesch C. 1995. Innovations in wild chimpanzees. *International Journal of Primatology* 16:1–16.

Bogin B. 2020. *Patterns of human growth*. 3rd ed. Cambridge, UK: Cambridge University Press.

Bogin B, Smith HB. 2012. Evolution of the human life cycle. In Stinson S, Bogin B, O'Rourke DH, eds., *Human Biology: An evolutionary and biocultural perspective*, 515–586. 2nd ed. Hoboken, NJ: Wiley.

Bolhuis J. 1991. Mechanisms of avian imprinting: A review. *Biological Review* 66:303–345.

Bolhuis JJ, Beckers GJ, Huybregts MA, Berwick RC, Everaert MB. 2018. Meaningful syntactic structure in songbird vocalizations? *PLoS Biology* 16(6):e2005157.

Bonardi C, Rey V, Richmond M, Hall G. 1993. Acquired equivalence of cues in pigeon autoshaping: Effects of training with common consequences and with common antecedents. *Animal Learning & Behavior* 21(4):369–376.

Bonnie KE, Horner V, Whiten A, de Wall FBM. 2007. Spread of arbitrary conventions among chimpanzees: A controlled experiment. *Proceedings Royal Society London B* 274:367–372.

Bornstein MH, Vibbert M, Tal J, O'Donnell K. 1992. Toddler language and play in the second year: Stability, covariation and influences of parenting. *First Language* 12(36):323–338.

Boserup E, Chambers R. 1965. *The conditions of agricultural growth: The economics of agrarian change under population pressure*. London: Routledge.

Bossema I. 1979. Jays and oaks: An eco-ethological study of a symbiosis. *Behaviour* 70(1–2): 1–117.

Bouchekioua Y, Blaisdell AP, Kosaki Y, Tsutsui-Kimura I, Craddock P, Mimura M, Watanabe S. 2021. Spatial inference without a cognitive map: The role of higher-order path integration. *Biological Reviews* 96(1):52–65.

Bouchekioua Y, Kosaki Y, Watanabe S, Blaisdell AP. 2021. Higher-order conditioning in the spatial domain. *Frontiers in Behavioral Neuroscience* 15:293.

Bouton ME. 2016. *Learning and behavior: A contemporary synthesis*. 2nd ed. Sunderland, MA: Sinauer.

Bowers RI, Timberlake W. 2018. Causal reasoning in rats' behaviour systems. *Royal Society Open Science* 5(7):171448.

Bowles S, Gintis H. 2011. *A cooperative species: Human reciprocity and its evolution*. Princeton, NJ: Princeton University Press.

Boyd R, Richerson PJ. 1982. Cultural transmission and the evolution of cooperative behavior. *Human Ecology* 10(3):325–351.

Boyd R, Richerson PJ. 1985. *Culture and the evolutionary process*. Chicago: University of Chicago Press.

Boyd R, Richerson PJ. 1996. Why culture is common, but cultural evolution is rare. *Proceedings of the British Academy* 88:77–93.

Boyd R, Richerson PJ. 2005. How microevolutionary processes give rise to history. In Boyd R, Richerson PJ, eds., *The origin and evolution of cultures*, 287–309. New York: Oxford University Press.

Boyer P, Wertsch JV, eds. 2009. *Memory in mind and culture.* Cambridge, UK: Cambridge University Press.

Bradbury JW, Vehrencamp SL. 2011. *Principles of animal communication.* 2nd ed. Sunderland, MA: Sinauer.

Braithwaite V. 2010. *Do fish feel pain?* New York: Oxford University Press.

Brandt AK, Slevc R, Gebrian M. 2012. Music and early language acquisition. *Frontiers in Psychology* 3:327.

Brannon EM. 2005. What animals know about numbers. In Campbell JID, ed., *Handbook of mathematical cognition,* 85–107. New York: Psychology Press.

Brass M, Heyes C. 2005. Imitation: Is cognitive neuroscience solving the correspondence problem? *Trends in Cognitive Sciences* 9(10):489–495.

Breland K, Breland M. 1961. The misbehavior of organisms. *American Psychologist* 61:681–684.

Brinums M, Imuta K, Suddendorf T. 2018. Practicing for the future: Deliberate practice in early childhood. *Child Development* 89(6):2051–2058.

Brockmann HJ. 1985. Tool use in digger wasps (Hymenoptera: Sphecinae). *Psyche* 92(2–3): 309–330.

Brown DE. 1991. *Human universals.* New York: McGraw-Hill.

Brozoski TJ, Brown RM, Rosvold H, Goldman PS. 1979. Cognitive deficit caused by regional depletion of dopamine in prefrontal cortex of rhesus monkey. *Science* 205:929–932.

Brumm A, Oktaviana AA, Burhan B, Hakim B, Lebe R, Zhao JX, Sulistyarto PH, Ririmasse M, Adhityatama S, Sumantri I, et al. 2021. Oldest cave art found in Sulawesi. *Science Advances* 7(3):eabd4648.

Bruner JS. 1966. *Toward a theory of instruction.* Cambridge, MA: Harvard University Press.

Bruner JS. 1972. Nature and uses of immaturity. *American Psychologist* 27(8):687.

Bruner JS. 1975. The ontogenesis of speech acts. *Journal of child language* 2(01):1–19.

Bryant PE, Bradley LL. 1985. *Rhyme and reason in reading and spelling.* Ann Arbor: University of Michigan Press.

Bryant PE, MacLean M, Bradley LL, Crossland J. 1990. Rhyme and alliteration, phoneme detection, and learning to read. *Developmental Psychology* 26(3):429.

Brzović Z. 2021. Natural kinds. In *The internet encyclopedia of philosophy.* https://iep.utm.edu.

Budaev S, Jørgensen C, Mangel M, Eliassen S, Giske J. 2019. Decision-making from the animal perspective: Bridging ecology and subjective cognition. *Frontiers in Ecology and Evolution* 7:164.

Bugnyar T, Huber L. 1997. Push or pull: An experimental study on imitation in marmosets. *Animal Behaviour* 54(4):817–831.

Buller DJ. 2006. *Adapting minds: Evolutionary psychology and the persistent quest for human nature.* Cambridge, MA: MIT Press.

Burkart JM, Schubiger MN, van Schaik CP. 2017. The evolution of general intelligence. *Behavioral and Brain Sciences* 40:E195.

Buskell A, Enquist M, Jansson F. 2019. A systems approach to cultural evolution. *Palgrave Communications* 5(1):1–15.

Busnel RG, Classe A. 2013. *Whistled languages.* Berlin: Springer.

Buss DM. 2000. *The dangerous passion: Why jealousy is as necessary as love or sex*. London: Bloomsbury.

Buss DM. 2005. *The handbook of evolutionary psychology*. Hoboken, NJ: John Wiley & Sons.

Butterworth B, Gallistel CR, Vallortigara G. 2018. Introduction: The origins of numerical abilities. *Philosophical Transactions of the Royal Society B* 373(1740):20160507.

Byrne RW. 1995. *The thinking ape: Evolutionary origins of intelligence*. New York: Oxford University Press.

Byrne RW, Whiten A. 1988. *Machiavellian intelligence: Social expertise and the evolution of intellect in monkeys, apes, and humans*. Oxford, UK: Clarendon Press.

Calcott B, Sterelny K. 2011. *The major transitions in evolution revisited*. Cambridge, MA: MIT Press.

Caldwell CA, Whiten A. 2004. Testing for social learning and imitation in common marmosets, *Callithrix jacchus*, using an artificial fruit. *Animal Cognition* 7(2):77–85.

Call J. 2012. Theory of mind in animals. In Seel NM, ed., *Encyclopedia of the sciences of learning*, 3316–3319. New York: Springer.

Call J, Carpenter M, Tomasello M. 2005. Copying results and copying actions in the process of social learning: Chimpanzees (*Pan troglodytes*) and human children (*Homo sapiens*). *Animal Cognition* 8:151–163.

Call J, Tomasello M. 1996. The effect of humans on the cognitive development of apes. In Russon AE, Bard KA, Parker ST, eds., *Reaching into thought: The minds of the great apes*, 371–403. Cambridge, UK: Cambridge University Press.

Campbell F, Heyes C, Goldsmith A. 1999. Stimulus learning and response learning by observation in the European starling, in a two-object/two-action test. *Animal Behaviour* 58(1):151–158.

Capaldi EJ. 1958. The effects of different amounts of training on the resistance to extinction of different patterns of partially reinforced responses. *Journal of Comparative and Physiological Psychology* 51(3):367.

Capaldi EJ. 1971. Memory and learning: A sequential viewpoint. In Honig WK, James PHR, eds., *Animal memory*, 111–154. New York: Academic Press.

Carey S. 2009. *The origin of concepts*. New York: Oxford University Press.

Carneiro RL. 2003. *Evolutionism in cultural anthropology*. New York: Routledge.

Caro T. 1994. *Cheetahs of the Serengeti Plains: Group living in an asocial species*. Chicago: University of Chicago Press.

Caro T, Hauser M. 1992. Is there teaching in nonhuman animals? *Quarterly Review of Biology* 67(2):151–174.

Case R. 1978. Intellectual development from birth to adulthood: A neo-Piagetian interpretation. In Siegler R, ed., *Children's thinking*, 37–72. Hillsdale, NJ: Lawrence Erlbaum.

Catchpole CK, Slater PJ. 2003. *Bird song: Biological themes and variations*. Cambridge, UK: Cambridge University Press.

Catmur C, Walsh V, Heyes C. 2009. Associative sequence learning: The role of experience in the development of imitation and the mirror system. *Philosophical Transactions of the Royal Society B* 364(1528):2369–2380.

Cavalli Sforza LL, Feldman MW. 1981. *Cultural transmission and evolution*. Princeton, NJ: Princeton University Press.

Cavalli Sforza LL, Menozzi P, Piazza A. 1996. *The history and geography of human genes.* Princeton, NJ: Princeton University Press.

Chabris CF, Hearst ES. 2003. Visualization, pattern recognition, and forward search: Effects of playing speed and sight of the position on grandmaster chess errors. *Cognitive Science* 27(4):637–648.

Chall J. 1979. The great debate: Ten years later, with a modest proposal for reading stages. In Resnick L, Weaver P, eds., *Theory and practice of early reading,* vol. 1, 29–55. Hillsdale, NJ: Lawrence Erlbaum.

Changizi MA. 2003. Relationship between number of muscles, behavioral repertoire size, and encephalization in mammals. *Journal of Theoretical Biology* 220:157–168.

Charman T, Baron-Cohen S, Swettenham J, Baird G, Cox A, Drew A. 2000. Testing joint attention, imitation, and play as infancy precursors to language and theory of mind. *Cognitive Development* 15(4):481–498.

Charness N, Tuffiash M, Krampe R, Reingold E, Vasyukova E. 2005. The role of deliberate practice in chess expertise. *Applied Cognitive Psychology* 19(2):151–165.

Chase WG, Simon HA. 1973. Perception in chess. *Cognitive Psychology* 4(1):55–81.

Cheke LG, Clayton NS. 2012. Eurasian jays (*Garrulus glandarius*) overcome their current desires to anticipate two distinct future needs and plan for them appropriately. *Biology Letters* 8(2):171–175.

Chen J, Van Rossum D, Ten Cate C. 2015. Artificial grammar learning in zebra finches and human adults: XYX versus XXY. *Animal Cognition* 18(1):151–164.

Chen JS, Amsel A. 1980. Recall (versus recognition) of taste and immunization against aversive taste anticipations based on illness. *Science* 209(4458):831–833.

Cheney DL, Seyfarth RM, Silk JB. 1995. The responses of female baboons (*Papio cynocephalus ursinus*) to anomalous social interactions: Evidence for causal reasoning? *Journal of Comparative Psychology* 109(2):134.

Chevalier-Skolnikoff S. 1989. Spontaneous tool use and sensorimotor intelligence in cebus compared with other monkeys and apes. *Behavioral and Brain Sciences* 12(03): 561–588.

Chevalier-Skolnikoff S, Liska J. 1993. Tool use by wild and captive elephants. *Animal Behaviour* 46(2):209–219.

Chi MT. 2006. Laboratory methods for assessing experts' and novices' knowledge. In Ericsson KA, Hoffman RR, Kozbelt A, Williams AM, eds., *The Cambridge handbook of expertise and expert performance,* 167–184. Cambridge, UK: Cambridge University Press.

Chi MT, Glaser R, Farr MJ. 2014. *The nature of expertise.* Hove, East Sussex, UK: Psychology Press.

Chiappe D, MacDonald K. 2005. The evolution of domain-general mechanisms in intelligence and learning. *Journal of General Psychology* 132(1):5–40.

Chivers DP, Smith RJF. 1994. Fathead minnows, *Pimephales promelas,* acquire predator recognition when alarm substance is associated with the sight of unfamiliar fish. *Animal Behaviour* 48(3):597–605.

Chomsky N. 1980. Rules and representations. *Behavioral and Brain Sciences* 3(01):1–15.

Christel M. 1993. Grasping techniques and hand preferences in Hominoidea. In Preuschoft H, Chivers DJ, eds., *Hands of primates,* 91–108. Vienna: Springer.

Christiansen MH, MacDonald MC. 2009. A usage-based approach to recursion in sentence processing. *Language Learning* 59:126–161.

Ciancia F. 1991. Tolman and Honzik (1930) revisited or the mazes of psychology (1930–1980). *Psychological Record* 41:461–472.

Clark EV. 1995. *Language acquisition: The lexicon and syntax*. New York: Academic Press.

Clark HB, Sherman JA. 1970. Effects of a conditioned reinforcer upon accuracy of match-to-sample behavior in pigeons. *Journal of the Experimental Analysis of Behavior* 13(3):375–384.

Clayton KN. 1969. Reward and reinforcement in selective learning: Considerations with respect to a mathematical model of learning. In Tapp JT, ed., *Reinforcement and behavior*, 96–119. New York: Academic Press.

Clayton NS, Bussey TJ, Dickinson A. 2003. Can animals recall the past and plan for the future? *Nature Reviews Neuroscience* 4(8):685–691.

Clayton NS, Dickinson A. 1998. Episodic-like memory during cache recovery by scrub jays. *Nature* 395(6699):272–274.

Clayton NS, Dickinson A. 1999a. Memory for the content of caches by scrub jays (*Aphelocoma coerulescens*). *Journal of Experimental Psychology: Animal Behavior Processes* 25(1):82–91.

Clayton NS, Dickinson A. 1999b. Scrub jays (*Aphelocoma coerulescens*) remember the relative time of caching as well as the location and content of their caches. *Journal of Comparative Psychology* 113(4):403–416.

Clayton NS, Farrar B, Boeckle M. 2020. Replications in comparative cognition: What should we expect and how can we improve? *Animal Behavior and Cognition* 7(1):1–22.

Clayton NS, Salwiczek LH, Dickinson A. 2007. Episodic memory. *Current Biology* 17(6):R189–R191.

Clutton-Brock T. 1991. *The evolution of parental care*. Princeton, NJ: Princeton University Press.

Clutton-Brock T. 2009. Cooperation between non-kin in animal societies. *Nature* 462(7269):51.

Colombo M, Eickhoff AE, Gross CG. 1993. The effects of inferior temporal and dorsolateral frontal lesions on serial-order behavior and visual imagery in monkeys. *Cognitive Brain Research* 1(4):211–217.

Colombo M, Scarf D. 2020. Are there differences in "intelligence" between nonhuman species? The role of contextual variables. *Frontiers in Psychology* 11:2072.

Colwill RM, Rescorla RA. 1985. Postconditioning devaluation of a reinforcer affects instrumental responding. *Journal of Experimental Psychology: Animal Behavior Processes* 11(1):120–132.

Conner DB, Knight DK, Cross DR. 1997. Mothers' and fathers' scaffolding of their 2-year-olds during problem-solving and literacy interactions. *British Journal of Developmental Psychology* 15(3):323–338.

Cook M, Mineka S. 1990. Selective associations in the observational conditioning of fear in rhesus monkeys. *Journal of Experimental Psychology: Animal Behavior Processes* 16(4):372–389.

Cook RG, Fowler C. 2014. "Insight" in pigeons: Absence of means-end processing in displacement tests. *Animal Cognition* 17(2):207–220.

Coolidge FL, Wynn T. 2005. Working memory, its executive functions, and the emergence of modern thinking. *Cambridge Archaeological Journal* 15(01):5–26.

Coradeschi S, Loutfi A, Wrede B. 2013. A short review of symbol grounding in robotic and intelligent systems. *KI-Künstliche Intelligenz* 27(2):129–136.

Corballis MC. 2011. *The recursive mind: The origins of human language, thought, and civilization.* Princeton, NJ: Princeton University Press.

Cormen TH, Leiserson CE, Rivest RL, Stein C. 2009. *Introduction to algorithms.* Cambridge, MA: MIT press.

Cornélio AM, de Bittencourt-Navarrete RE, de Bittencourt Brum R, Queiroz CM, Costa MR. 2016. Human brain expansion during evolution is independent of fire control and cooking. *Frontiers in Neuroscience* 10:167.

Correia SP, Dickinson A, Clayton NS. 2007. Western scrub-jays anticipate future needs independently of their current motivational state. *Current Biology* 17(10):856–861.

Cosmides L, Barrett HC, Tooby J. 2010. Adaptive specializations, social exchange, and the evolution of human intelligence. *Proceedings of the National Academy of Sciences* 107(2):9007–9014.

Cosmides L, Tooby J. 1992. Cognitive adaptations for social exchange. In Barkow JH, Cosmides L, Tooby J, eds., *The adapted mind,* 163–228. New York: Oxford University Press.

Cosmides L, Tooby J. 1994. Origins in domain specificity: The evolution of functional organization. In Hirschfeld LA, Gelman SA, eds., *Mapping the mind: Domain specificity in cognition and culture.* Cambridge, UK: Cambridge University Press.

Cosmides L, Tooby J, Barkow JH. 1992. Introduction: Evolutionary psychology and conceptual integration. In Barkow JH, Cosmides L, Tooby J, eds., *The adapted mind.* New York: Oxford University Press.

Cowan N, Hulmes C, eds. 1997. *The development of memory in childhood.* Hove, East Sussex, UK: Psychology Press.

Craik K. 1943. *The Nature of Explanation.* Cambridge, UK: Cambridge University Press.

Creanza N, Kolodny O, Feldman MW. 2017. Greater than the sum of its parts? Modelling population contact and interaction of cultural repertoires. *Journal of the Royal Society Interface* 14(130):20170171.

Cronin PB. 1980. Reinstatement of postresponse stimuli prior to reward in delayed-reward discrimination learning by pigeons. *Animal Learning & Behavior* 8(3):352–358.

Crystal JD. 2010. Episodic-like memory in animals. *Behavioural Brain Research* 215(2):235–243.

Csibra G, Gergely G. 2007. "Obsessed with goals": Functions and mechanisms of teleological interpretation of actions in humans. *Acta Psychologica* 124(1):60–78.

Cubek R, Ertel W, Palm G. 2015. A critical review on the symbol grounding problem as an issue of autonomous agents. In *Joint German/Austrian Conference on Artificial Intelligence (Künstliche Intelligenz),* 256–263. Cham, Switzerland: Springer.

Curio E. 1988. Cultural transmission of enemy recognition by birds. In Zentall T, Galef BG Jr., eds., *Social learning,* 75–98. Hillsdale, NJ: Lawrence Erlbaum.

Curio E, Ernst U, Vieth W. 1978. Cultural transmission of enemy recognition: One function of mobbing. *Science* 202:899–901.

Curtiss S. 1977. *Genie: A psycholinguistic study of a modern-day wild child.* Boston: Academic Press.

Custance D, Prato-Previde E, Spiezio C, Rigamonti MM, Poli M. 2006. Social learning in pig-tailed macaques (*Macaca nemestrina*) and adult humans (*Homo sapiens*) on a two-action artificial fruit. *Journal of Comparative Psychology* 120(3):303.

Cutler RG. 1976. Evolution of longevity in primates. *Journal of Human Evolution* 5(2):169–202.

Dall SRX, Cuthill IC. 1997. The information costs of generalism. *Oikos* 80(1): 197–202.

Daly M, Wilson M. 1988. *Homicide*. New York: Routledge.

Daly M, Wilson M. 1998. *The truth about Cinderella: A Darwinian view of parental love*. New Haven, CT: Yale University Press.

Daniels PT, Bright W. 1996. *The world's writing systems*. New York: Oxford University Press.

Dart RA. 1925. *Australopithecus africanus*, the man-ape of South Africa. *Nature* 115:195–199.

Davis H, Pérusse R. 1988. Numerical competence in animals: Definitional issues, current evidence, and a new research agenda. *Behavioral and Brain Sciences* 11(4):561–579.

Deacon RM, Rawlins JNP. 2006. T-maze alternation in the rodent. *Nature Protocols* 1(1):7.

Deacon TW. 1997. *The symbolic species: The co-evolution of language and the brain*. New York: W. W. Norton.

Dean LG, Kendal RL, Schapiro SJ, Thierry B, Laland KN. 2012. Identification of the social and cognitive processes underlying human cumulative culture. *Science* 335:1114–1118.

Dean LG, Vale GL, Laland KN, Flynn E, Kendal RL. 2014. Human cumulative culture: A comparative perspective. *Biological Reviews* 89(2):284–301.

Deary IJ, Penke L, Johnson W. 2010. The neuroscience of human intelligence differences. *Nature Reviews Neuroscience* 11(3):201.

Deary IJ, Spinath FM, Bates TC. 2006. Genetics of intelligence. *European Journal of Human Genetics* 14(6):690–700.

Dehaene S. 2009. Origins of mathematical intuitions. *Annals of the New York Academy of Sciences* 1156(1):232–259.

Dekleva M, Dufour V, De Vries H, Spruijt BM, Sterck EH. 2011. Chimpanzees (*Pan troglodytes*) fail a what-where-when task but find rewards by using a location-based association strategy. *PLoS One* 6(2):e16593.

Delius J. 1994. Comparative cognition of identity. In Bertelson P, Eelen P, Ydewalle G, eds., *International Perspectives on Psychological Science*, 25–39. Hillsdale, NJ: Lawrence Erlbaum.

Delius J, Habers G. 1978. Symmetry: Can pigeons conceptualize it? *Behavioral Biology* 22:336–342.

Delius J, Nowak B. 1982. Visual symmetry recognition by pigeons. *Psychological Research* 44:199–212.

DeLoache JS, LoBue V. 2009. The narrow fellow in the grass: Human infants associate snakes and fear. *Developmental Science* 12(1):201–207.

de Montpellier G. 1933. An experiment on the order of elimination of blind alleys in maze learning. *Pedagogical Seminary and Journal of Genetic Psychology* 43(1):123–139.

Dempster FN. 1981. Memory span: Sources of individual and developmental differences. *Psychological Bulletin* 89(1):63.

Dennett D. 2009. The cultural evolution of words and other thinking tools. *Cold Spring Harbor Symposium in Quantitative Biology* 74:435–441.

De Resende BD, Ottoni EB, Fragaszy DM. 2008. Ontogeny of manipulative behavior and nut-cracking in young tufted capuchin monkeys (*Cebus apella*): A perception-action perspective. *Developmental Science* 11(6):828–840.

Deutsch J, Moore B, Heinrichs S. 1989. Unlearned specific appetite for protein. *Physiology & Behavior* 46(4):619–624.

Dickinson A. 1980. *Contemporary animal learning theory*. Cambridge, UK: Cambridge University Press.

Dickinson A. 2012. Associative learning and animal cognition. *Philosophical Transactions of the Royal Society B* 367(1603):2733–2742.

Dickinson A, Burke J. 1996. Within-compound associations mediate retrospective revaluation of causality judgements. *Quarterly Journal of Experimental Psychology* 49B:60–80.

Dickinson A, Wood N, Smith JW. 2002. Alcohol seeking by rats: Action or habit? *Quarterly Journal of Experimental Psychology: Section B* 55(4):331–348.

Dijkstra EW. 1952. A note on two problems in connexion with graphs. *Numerische Mathematik* 1(1):269–271.

Dinsmoor JA, Browne MP, Lawrence CE. 1972. A test of the negative discriminative stimulus as a reinforcer of observing. *Journal of the Experimental Analysis of Behavior* 18:79–85.

Dobzhansky T. 1964. Evolution organic and superorganic. *Bulletin of the Atomic Scientists* 20: 4–8.

Dolhinow P, Sarich V. 1971. *Background for man: Readings in physical anthropology*. Boston: Little, Brown.

Dollard J, Miller NE, Doob LW, Mowrer OH, Sears RR. 1939. *Frustration and aggression*. New Haven, CT: Yale University Press.

Domjan M. 1992. Adult learning and mate choice: Possibilities and experimental evidence. *American Zoologist* 32:48–61.

Domjan M. 2012. Biological or evolutionary constraints on learning. In Seel NM, ed., *Encyclopedia of the sciences of learning*, 461–463. Berlin: Springer.

Donald M. 1991. *Origins of the modern mind: Three stages in the evolution of culture and cognition*. Cambridge, MA: Harvard University Press.

Dosher B, Lu ZL. 2017. Visual perceptual learning and models. *Annual Review of Vision Science* 3:343–363.

Ducoing AM, Thierry B. 2005. Tool-use learning in Tonkean macaques (*Macaca tonkeana*). *Animal Cognition* 8(2):103–113.

D'Udine B, Alleva E. 1983. Early experience and sexual preferences in rodents. In Bateson P, ed., *Mate choice*, 311–327. Cambridge, UK: Cambridge University Press.

Dugatkin LA. 1997. *Cooperation among animals: An evolutionary perspective*. New York: Oxford University Press.

Dukas R, ed. 1998. *Cognitive ecology: The evolutionary ecology of information processing and decision making*. Chicago: University of Chicago Press.

Dunbar R. 1998. The social brain hypothesis. *Evolutionary Anthropology* 6:178–190.

Dunbar R, McAdam M, O'Connell S. 2005. Mental rehearsal in great apes (*Pan troglodytes* and *Pongo pygmaeus*) and children. *Behavioural Processes* 69(3):323–330.

Durham WH. 1991. *Coevolution: Genes, culture and human diversity*. Stanford, CA: Stanford University Press.

Dusenbery DB. 1992. *Sensory ecology: How organisms acquire and respond to information*. New York: W. H. Freeman.

Eibl Eibesfeldt I. 1963. Angeborenes und erworbenes im verhalten einiger säuger. *Zeitschrift für Tierpsychologie* 20:705–754.

Eibl Eibesfeldt I. 1975. *Ethology. The biology of behavior*. New York: Holt, Rinehart and Winston.

Eichenbaum H, Fagan A, Cohen N. 1986. Normal olfactory discrimination learning set and facilitation of reversal learning after medial-temporal damage in rats: Implications for an account of preserved learning abilities in amnesia. *Journal of Neuroscience* 6(7):1876–1884.

Ellen P, Pate JL. 1986. Is insight merely response chaining? A reply to Epstein. *Psychological Record* 36(2):155–160.

Ellis D, Brunson S. 1993. Tool use by the red-tailed hawk (*Buteo jamaicensis*). *Journal of Raptor Research* 27(2):128.

Ellis N, Large B. 1988. The early stages of reading: A longitudinal study. *Applied Cognitive Psychology* 2(1):47–76.

Elner RW, Hughes RN. 1978. Energy maximization in the diet of the shore crab, *Carcinus maenas*. *Journal of Animal Ecology* 47:103–116.

Emery NJ, Clayton NS. 2009. Tool use and physical cognition in birds and mammals. *Current Opinion in Neurobiology* 19(1):27–33.

Emery NJ, Clayton NS, Frith CD. 2008. *Social intelligence: From brain to culture*. Oxford, UK: Oxford University Press.

Enquist M, Ghirlanda S. 2005. *Neural networks and animal behavior*. Princeton, NJ: Princeton University Press.

Enquist M, Ghirlanda S. 2007. Evolution of social learning does not explain the origin of human cumulative culture. *Journal of Theoretical Biology* 246:129–135.

Enquist M, Ghirlanda S, Eriksson K. 2011. Modeling the evolution and diversity of complex cumulative culture. *Philosophical Transactions of the Royal Society B* 366:412–423.

Enquist M, Ghirlanda S, Jarrick A, Wachtmeister CA. 2008. Why does human culture increase exponentially? *Theoretical Population Biology* 74:46–55.

Enquist M, Lind J, Ghirlanda S. 2016. The power of associative learning and the ontogeny of optimal behaviour. *Royal Society Open Science* 3(11):160734.

Epstein R. 1981. On pigeons and people: A preliminary look at the Columban Simulation Project. *Behavior Analyst* 4(1):43–55.

Epstein R. 1985. The spontaneous interconnection of three repertoires. *Psychological Record* 35(2):131–141.

Epstein R. 1987. The spontaneous interconnection of four repertoires of behavior in a pigeon (*Columba livia*). *Journal of Comparative Psychology* 101(2):197.

Epstein R, Kirshnit CE, Lanza RP, Rubin LC. 1984. "Insight" in the pigeon: Antecedents and determinants of an intelligent performance. *Nature* 308:61–62.

Epstein R, Lanza RP, Skinner BF. 1980. Symbolic communication between two pigeons (*Columba livia domestica*). *Science* 207(4430):543–545.

Epstein R, Medalie SD. 1983. The spontaneous use of a tool by a pigeon. *Behaviour Analysis Letters* 3(4):241–247.

Erard M. 2012. *Babel no more: The search for the world's most extraordinary language learners.* New York: Simon & Schuster.

Ericsson KA. 2006. The influence of experience and deliberate practice on the development of superior expert performance. In Ericsson KA, Charness N, Feltovich PJ, Hoffman RR, eds., *The Cambridge handbook of expertise and expert performance*, vol. 38, 685–705. Cambridge, UK: Cambridge University Press.

Ericsson KA, Hoffman RR, Kozbelt A, Williams AM., eds. 2018. *The Cambridge handbook of expertise and expert performance.* Cambridge, UK: Cambridge University Press.

Ericsson KA, Krampe RT, Tesch-Römer C. 1992. The role of deliberate practice in the acquisition of expert performance. *Psychological Review* 100(3):363.

Eriksson K, Enquist M, Ghirlanda S. 2007. Critical points in current theory of conformist social learning. *Journal of Evolutionary Psychology* 5:67–88.

Espinosa JS, Stryker MP. 2012. Development and plasticity of the primary visual cortex. *Neuron* 75(2):230–249.

Evans S. 1936. Flexibility of established habit. *Journal of General Psychology* 14(1):177–200.

Evans TA, Westergaard GC. 2006. Self-control and tool use in tufted capuchin monkeys (*Cebus apella*). *Journal of Comparative Psychology* 120(2):163.

Ewer RF. 1968. *Ethology of mammals.* London: Logos Press.

Ewer RF. 1969. The "instinct to teach." *Nature* 222:698.

Ewert JP. 1980. *Neuroethology.* Berlin: Springer-Verlag.

Fagot J, Cook RG. 2006. Evidence for large long-term memory capacities in baboons and pigeons and its implications for learning and the evolution of cognition. *Proceedings of the National Academy of Sciences* 103(46):17564–17567.

Fantino E. 1977. Conditioned reinforcement: Choice and information. In Honig WK, Staddon JER, eds., *Handbook of operant behavior*, 313–339. Englewood Cliffs, NJ: Prentice Hall.

Fantino E, Logan C. 1979. *The experimental analysis of behavior: A biological perspective.* San Francisco: W. H. Freeman.

Farrar BG, Voudouris K, Clayton NS. 2021. Replications, comparisons, sampling and the problem of representativeness in animal cognition research. *Animal Behavior and Cognition* 8(2):273.

Farris HE. 1967. Classical conditioning of courting behavior in the Japanese quail, *Coturnix coturnix japonica. Journal of the Experimental Analysis of Behavior* 10(2):213.

Fehr E, Gächter S. 2000. Fairness and retaliation: The economics of reciprocity. *Journal of Economic Perspectives* 14(3):159–181.

Feigenbaum EA. 1977. The art of artificial intelligence. 1. Themes and case studies of knowledge engineering. Technical report, Stanford University Department of Computer Science.

Feldman MW, Aoki K, Kumm J. 1996. Individual versus social learning: Evolutionary analysis in a fluctuating environment. *Anthropological Science* 104:209–213.

Ferkin MH, Combs A, Pierce AA, Franklin S. 2008. Meadow voles, *Microtus pennsylvanicus*, have the capacity to recall the "what," "where," and "when" of a single past event. *Animal Cognition* 11(1):147–159.

Festinger L. 1957. *A theory of cognitive dissonance.* Stanford, CA: Stanford University Press.

Fimi D, Higgins A. 2016. *A secret vice: Tolkien on invented languages.* New York: HarperCollins.

Fiorito G, Scotto P. 1992. Observational learning in *Octopus vulgaris*. *Science* 256(5056):545–547.

Fischer KW, Bidell TR. 2007. Dynamic development of action and thought. In Damon W, Lerner RM, eds., *Handbook of child psychology*, 313–399. 6th ed. New York: Wiley.

Fishbein AR, Idsardi WJ, Ball GF, Dooling RJ. 2020. Sound sequences in birdsong: How much do birds really care? *Philosophical Transactions of the Royal Society B* 375(1789): 20190044.

Fisher J, Hinde RA. 1949. The opening of milk bottles by birds. *British Birds* XLII:347–357.

Fivush R. 2014. Emotional content of parent-child conversations about the past. In Nelson CA, ed., *Memory and affect in development*, vol. 26, 39–77. Mahwah, NJ: Lawrence Erlbaum.

Fivush R, Hudson JA, eds. 1990. *Knowing and remembering in young children*. Cambridge, UK: Cambridge University Press.

Flavell J. 1976. Metacognitive aspects of problem solving. In Resnick LB, ed., *The nature of intelligence*, 333–335. Hillsdale, NJ: Lawrence Erlbaum.

Fodor JA. 1975. *The language of thought*. Cambridge, MA: Harvard University Press.

Fodor JA. 1983. *The modularity of mind*. Cambridge, MA: MIT Press.

Fogarty L, Creanza N, Feldman M. 2013. The role of cultural transmission in human demographic change: An age-structured model. *Theoretical Population Biology* 88:68–77.

Foree DD, LoLordo VM. 1973. Attention in the pigeon: Differential effects of food-getting versus shock-avoidance procedures. *Journal of Comparative and Physiological Psychology* 85(3):551.

Fountain SB. 2008. Pattern structure and rule induction in sequential learning. *Comparative Cognition & Behavior Reviews* 3:66–85.

Fountain SB, Rowan JD. 1995. Coding of hierarchical versus linear pattern structure in rats and humans. *Journal of Experimental Psychology: Animal Behavior Processes* 21(3):187.

Fox M. 1969. Ontogeny of prey-killing behavior in Canidae. *Behaviour* 35(3–4):259–272.

Frank MC, Everett DL, Fedorenko E, Gibson E. 2008. Number as a cognitive technology: Evidence from Pirahã language and cognition. *Cognition* 108(3):819–824.

Frankenhuis WE, Panchanathan K, Barto AG. 2019. Enriching behavioral ecology with reinforcement learning methods. *Behavioural Processes* 161:64–100.

Frankenhuis WE, Walasek N. 2020. Modeling the evolution of sensitive periods. *Developmental Cognitive Neuroscience* 41:100715.

Fraser KM, Holland PC. 2019. Occasion setting. *Behavioral Neuroscience* 133(2):145.

Friedgut E, Kalai G. 1996. Every monotone graph property has a sharp threshold. *Proceedings of the American Mathematical Society* 124(10):2993–3002.

Friedman SL, Scholnick EK, eds. 1997. *The developmental psychology of planning: Why, how, and when do we plan?* New York: Psychology Press.

Frieze A, Karoński M. 2016. *Introduction to random graphs*. Cambridge, UK: Cambridge University Press.

Frith CD, Frith U. 2012. Mechanisms of social cognition. *Annual Review of Psychology* 63:287–313.

Frith U, Frith CD. 2010. The social brain: Allowing humans to boldly go where no other species has been. *Philosophical Transactions of the Royal Society B* 365(1537):165–176.

Fugazza C, Miklósi Á. 2015. Social learning in dog training: The effectiveness of the do as I do method compared to shaping/clicker training. *Applied Animal Behaviour Science* 171:146–151.

Fugazza C, Pogány Á, Miklósi Á. 2016. Do as I ... did! Long-term memory of imitative actions in dogs (*Canis familiaris*). *Animal Cognition* 19(2):263–269.

Furuichi T, Sanz C, Koops K, Sakamaki T, Ryu H, Tokuyama N, Morgan D. 2015. Why do wild bonobos not use tools like chimpanzees do? *Behaviour* 152:425–460.

Gabora L. 1997. The origin and evolution of culture and creativity. *Journal of Memetics* 1(1): 1–28.

Gabora L. 1998. Autocatalytic closure in a cognitive system: A tentative scenario for the origin of culture. *Psycholoquy* 9(67):1–26.

Gabora L, Steel M. 2017. Autocatalytic networks in cognition and the origin of culture. *Journal of Theoretical Biology* 431:87–95.

Gabora L, Steel M. 2020. Modeling a cognitive transition at the origin of cultural evolution using autocatalytic networks. *Cognitive Science* 44(9):e12878.

Galef BG Jr. 1988. Imitations in animals: History, definitions, and interpretations of data from the psychological laboratory. In Zentall T, Galef BG Jr., eds., *Social learning*, vol. 1, 3–28. Mahwah, NJ: Lawrence Erlbaum.

Galef BG Jr. 1996. Social enhancement of food preferences in Norway rats: A brief review. In Heyes C, Galef BG Jr., eds., *Social learning in animals: The roots of culture*. New York: Academic Press.

Galef BG Jr., Manzig L, Field R. 1986. Imitation learning in budgerigars: Dawson and Foss (1965) revisited. *Behavioural Processes* 13(1):191–202.

Gallistel CR. 1990. *The organization of learning*. Cambridge, MA: MIT Press.

Gallistel CR. 1999. Reinforcement learning. *Journal of Cognitive Neuroscience* 11(1):126–134.

Garcia J, Ervin FA, Koelling RA. 1966. Learning with prolonged delay of reinforcement. *Psychonomic Science* 5:121–122.

Garcia J, Ervin FA, Koelling RA. 1967. Trace conditioning with X-rays as the aversive stimulus. *Psychonomic Science* 9:11–12.

Garcia J, Koelling RA. 1966. Relation of cue to consequence in avoidance learning. *Psychonomic Science* 4:123–124.

Gärdenfors P. 2006. *How Homo became sapiens: On the evolution of thinking*. New York: Oxford University Press.

Gardner H. 2011. *Frames of mind: The theory of multiple intelligences*. New York: Basic Books.

Gauvain M. 2001. *The social context of cognitive development*. New York: Guilford Press.

Gauvain M, Huard RD. 1999. Family interaction, parenting style, and the development of planning: A longitudinal analysis using archival data. *Journal of Family Psychology* 13(1):75.

Gauvain M, Perez SM, Reisz Z. 2018. Stability and change in mother-child planning over middle childhood. *Developmental Psychology* 54(3):571.

Gauvain M, Rogoff B. 1989. Collaborative problem solving and children's planning skills. *Developmental Psychology* 25(1):139.

Geertz C. 1973. *The interpretation of cultures*. New York: Basic Books.

Gegenfurtner KR, Sharpe LT. 1999. *Color vision*. Cambridge, UK: Cambridge University Press.

Gelman R, Williams EM. 1998. Enabling constraints for cognitive development and learning: Domain specificity and epigenesis. In Damon W, ed., *Handbook of child psychology: Cognition, perception, and language*, vol. 2, 575–630. New York: Wiley & Sons.

Gelman SA, Roberts SO. 2017. How language shapes the cultural inheritance of categories. *Proceedings of the National Academy of Sciences* 114(30):7900–7907.

Gentner TQ, Fenn KM, Margoliash D, Nusbaum HC. 2006. Recursive syntactic pattern learning by songbirds. *Nature* 440(7088):1204–1207.

Gershman SJ, Blei DM, Niv Y. 2010. Context, learning, and extinction. *Psychological Review* 117(1):197.

Gertner Y, Fisher C, Eisengart J. 2006. Learning words and rules: Abstract knowledge of word order in early sentence comprehension. *Psychological Science* 17(8):684–691.

Ghirlanda S. 2015. On elemental and configural models of associative learning. *Journal of Mathematical Psychology* 64:8–16.

Ghirlanda S, Acerbi A, Herzog HA, Serpell JA. 2013. Fashion vs. function in cultural evolution: The case of dog breed popularity. *PLoS One* 8(9):e74770.

Ghirlanda S, Enquist M. 2003. A century of generalization. *Animal Behaviour* 66:15–36.

Ghirlanda S, Enquist M, Lind J. 2013. Coevolution of intelligence, behavioral repertoire, and lifespan. *Theoretical Population Biology* 91:44–49.

Ghirlanda S, Lind J. 2017. "Aesop's fable" experiments demonstrate trial-and-error learning in birds, but no causal understanding. *Animal Behaviour* 123:239–247.

Ghirlanda S, Lind J, Enquist M. 2017. Memory for stimulus sequences: A divide between humans and other animals? *Royal Society Open Science* 4(6):161011.

Ghirlanda S, Lind J, Enquist M. 2020. A-learning: A new formulation of associative learning theory. *Psychonomic Bulletin & Review* 27(6):1166–1194.

Gibson EJ. 1969. *Principles of perceptual learning and development*. New York: Appleton-Century-Crofts.

Gibson EJ, Walk RD. 1956. The effect of prolonged exposure to visually presented patterns on learning to discriminate between them. *Journal of Comparative and Physiological Psychology* 49:239–242.

Gibson KR. 1999. Social transmission of facts and skills in the human species: Neural mechanisms. 1999. In Box HO, Gibson HC, eds., *Mammalian social learning: Comparative and ecological perspectives*, 351–366. Cambridge, UK: Cambridge University Press.

Gilbert DT, Wilson TD. 2007. Prospection: Experiencing the future. *Science* 317(5843):1351–1354.

Gintis H, Bowles S, Boyd R, Fehr E. 2003. Explaining altruistic behavior in humans. *Evolution and Human Behavior* 24:153–172.

Giurfa M, Eichmann B, Menzel R. 1996. Symmetry perception in insects. *Nature* 382(6590): 458–461.

Giurfa M, Zhang S, Jenett A, Menzel R, Srinivasan MV. 2001. The concepts of "sameness" and "difference" in an insect. *Nature* 410(6831):930–933.

Glickman SE, Sroges RW. 1966. Curiosity in zoo animals. *Behaviour* 26(1/2):151–188.

Gobet F, Campitelli G. 2007. The role of domain-specific practice, handedness, and starting age in chess. *Developmental Psychology* 43(1):159.

Gobet F, Charness N. 2006. Expertise in chess. In Ericsson KA, Charness N, Feltovich PJ, Hoffman RR, eds., *The Cambridge handbook of expertise and expert performance*, 523–538. Cambridge, UK: Cambridge University Press.

Gobet F, Simon HA. 1996a. Recall of rapidly presented random chess positions is a function of skill. *Psychonomic Bulletin & Review* 3(2):159–163.

Gobet F, Simon HA. 1996b. Templates in chess memory: A mechanism for recalling several boards. *Cognitive Psychology* 31(1):1–40.

Godfrey-Smith P. 2001. Environmental complexity and the evolution of cognition. In Sternberg R, Kaufman J, eds., *The evolution of intelligence*, 233–249. Mahwah, NJ: Lawrence Erlbaum.

Goldstone RL. 1998. Perceptual learning. *Annual Review of Psychology* 49(1):585–612.

Gombrich EH. 1977. *Art & illusion: A study in the psychology of pictoral representation*. 5th ed. London: Phaidon.

González-Gómez PL, Bozinovic F, Vásquez RA. 2011. Elements of episodic-like memory in free-living hummingbirds, energetic consequences. *Animal Behaviour* 81(6):1257–1262.

Gopnik A, Frankenhuis W, Tomasello M. 2020. Life history and learning: How childhood, caregiving and old age shape cognition and culture in humans and other animals. *Philosophical Transactions of the Royal Society B* 375:1–151.

Gopnik A, Meltzoff AN, Bryant P. 1997. *Words, thoughts, and theories*. Cambridge, MA: MIT Press.

Gopnik A, Meltzoff AN, Kuhl PK. 1999. *The scientist in the crib: Minds, brains, and how children learn*. New York: Morrow.

Gopnik A, Sobel DM, Schulz LE, Glymour C. 2001. Causal learning mechanisms in very young children: Two-, three-, and four-year-olds infer causal relations from patterns of variation and covariation. *Developmental Psychology* 37(5):620.

Gordon P. 2004. Numerical cognition without words: Evidence from Amazonia. *Science* 306(5695):496–499.

Goss-Custard JD. 1977. Optimal foraging and the size selection of worms by redshank, *Tringa totanus*, in the field. *Animal Behaviour* 25:10–29.

Goto K, Watanabe S. 2009. Visual working memory of jungle crows (*Corvus macrorhynchos*) in operant delayed matching-to-sample. *Japanese Psychological Research* 51(3):122–131.

Green RE, Krause J, Briggs AW, Maricic T, Stenzel U, Kircher M, Patterson N, Li H, Zhai W, Fritz MHY, et al. 2010. A draft sequence of the Neandertal genome. *Science* 328(5979):710–722.

Greenberg DL, Verfaellie M. 2010. Interdependence of episodic and semantic memory: Evidence from neuropsychology. *Journal of the International Neuropsychological Society: JINS* 16(5):748.

Grice GR. 1948. The relationship of secondary reinforcement to delayed reward in visual discrimination training. *Journal of Experimental Psychology* 38:1–16.

Griebel U, Oller DK. 2012. Vocabulary learning in a Yorkshire terrier: Slow mapping of spoken words. *PLoS One* 7(2):e30182.

Griffiths D, Dickinson A, Clayton NS. 1999. Episodic memory: What can animals remember about their past? *Trends in Cognitive Sciences* 3(2):74–80.

Grosch J, Neuringer A. 1981. Self-control in pigeons under the Mischel paradigm. *Journal of the Experimental Analysis of Behavior* 35(1):3–21.

Guilford JP. 1967. *The nature of human intelligence.* New York: McGraw-Hill.

Guthrie ER. 1935. *The psychology of learning.* New York: Harper.

Guttman N. 1959. Generalization gradients around stimuli associated with different reinforcement schedules. *Journal of Experimental Psychology* 58(5):335–340.

Haden CA, Ornstein PA. 2009. Research on talking about the past: The past, present, and future. *Journal of Cognition and Development* 10(3):135–142.

Hall G. 1991. *Perceptual and associative learning.* Cambridge, UK: Cambridge University Press.

Hall G. 2002. Associative structures in Pavlovian and instrumental conditioning. In Pashler H, ed., *Stevens' handbook of experimental psychology.* New York: Wiley & Sons.

Halliday MAK. 1973. *Explorations in the functions of language.* London: Edward Arnold.

Hambrick DZ, Oswald FL, Altmann EM, Meinz EJ, Gobet F, Campitelli G. 2014. Deliberate practice: Is that all it takes to become an expert? *Intelligence* 45:34–45.

Hampton RR. 2001. Rhesus monkeys know when they remember. *Proceedings of the National Academy of Sciences* 98(9):5359–5362.

Hampton RR. 2019. Parallel overinterpretation of behavior of apes and corvids. *Learning and Behavior* 47(2):105–106.

Hanson H. 1959. Effects of discrimination training on stimulus generalization. *Journal of Experimental Psychology* 58(5):321–333.

Harley HE, Roitblat H, Nachtigall P. 1996. Object representation in the bottlenose dolphin (*Tursiops truncatus*): Integration of visual and echoic information. *Journal of Experimental Psychology: Animal Behavior Processes* 22(2):164.

Harmand S, Lewis JE, Feibel CS, Lepre CJ, Prat S, Lenoble A, Boës X, Quinn RL, Brenet M, Arroyo A, et al. 2015. 3.3-million-year-old stone tools from Lomekwi 3, West Turkana, Kenya. *Nature* 521(7552):310–315.

Harnad S. 2017. To cognize is to categorize: Cognition is categorization. In Cohen H, Lefebvre C, eds., *Handbook of categorization in cognitive science*, 21–54. 2nd ed. Amsterdam: Elsevier.

Harris M. 1999. *Theories of culture in postmodern times.* Walnut Creek, CA: Altamira Press.

Hart B, Risley TR. 1999. *The social world of children: Learning to talk.* Baltimore, MD: Paul H. Brookes.

Harvey PH, Clutton-Brock TH. 1985. Life history variation in primates. *Evolution* 39(3): 559–581.

Haselgrove M. 2016. Overcoming associative learning. *Journal of Comparative Psychology* 130(3):226.

Haselgrove M, Hogarth L. 2013. *Clinical applications of learning theory.* Hove, East Sussex, UK: Psychology Press.

Haskell M, Coerse NC, Forkman B. 2000. Frustration-induced aggression in the domestic hen: The effect of thwarting access to food and water on aggressive responses and subsequent approach tendencies. *Behaviour* 137(4):531–546.

Hassenstein B. 1971. *Information and control in the living organism: An elementary introduction.* London: Chapman & Hall.

Hattie J. 2008. *Visible learning: A synthesis of over 800 meta-analyses relating to achievement.* New York: Routledge.

Hattie J, Clarke S. 2018. *Visible learning: Feedback*. New York: Routledge.
Hauser M, Chomsky N, Fitch WT. 2002. The faculty of language: What is it, who has it, and how did it evolve? *Science* 298:1569–1579.
Hayden B. 1993. The cultural capacities of Neandertals: A review and re-evaluation. *Journal of Human Evolution* 24(2):113–146.
Hayes-Roth B. 1977. Evolution of cognitive structures and processes. *Psychological Review* 84(3):260.
Haykin S. 2008. *Neural networks and learning machines*. 3rd ed. Englewood Cliffs, NJ: Prentice Hall.
Healy S, Hurly T. 1995. Spatial memory in rufous hummingbirds (*Selasphorus rufus*): A field test. *Animal Learning & Behavior* 23(1):63–68.
Heckhausen J. 1987. Balancing for weaknesses and challenging developmental potential: A longitudinal study of mother-infant dyads in apprenticeship interactions. *Developmental Psychology* 23(6):762.
Heise GA, Conner R, Martin RA. 1976. Effects of scopolamine on variable intertrial interval spatial alternation and memory in the rat. *Psychopharmacology* 49(2):131–137.
Heise GA, Keller C, Khavari K, Laughlin N. 1969. Discrete-trial alternation in the rat. *Journal of the Experimental Analysis of Behavior* 12(4):609–622.
Heldstab SA, Isler K, Schuppli C, van Schaik CP. 2020. When ontogeny recapitulates phylogeny: Fixed neurodevelopmental sequence of manipulative skills among primates. *Science Advances* 6(30):eabb4685.
Helfman GS, Schultz ET. 1984. Social transmission of behavioural traditions in a coral reef fish. *Animal Behaviour* 32:379–384.
Henderson J, Hurly TA, Healy SD. 2001. Rufous hummingbirds' memory for flower location. *Animal Behaviour* 61(5):981–986.
Hendry DP, ed. 1969. *Conditioned reinforcement*. Homewood, IL: Dorsey Press.
Henneberg M. 1988. Decrease of human skull size in the Holocene. *Human Biology* 60(3): 395–405.
Hennefield L, Hwang HG, Weston SJ, Povinelli DJ. 2018. Meta-analytic techniques reveal that corvid causal reasoning in the Aesop's fable paradigm is driven by trial-and-error learning. *Animal Cognition* 21(6):735–748.
Henrich J. 2004. Demography and cultural evolution: How adaptive cultural processes can produce maladaptive losses—the Tasmanian case. *American Antiquity* 69(2):197–214.
Henrich J. 2015. *The secret of our success: How culture is driving human evolution, domesticating our species, and making us smarter*. Princeton, NJ: Princeton University Press.
Henrich J, Gil White F. 2001. The evolution of prestige: Freely conferred deference as a mechanism for enhancing the benefits of cultural transmission. *Evolution and Human Behavior* 22:65–96.
Henrich J, McElreath R. 2007. Dual inheritance theory: The evolution of human cultural capacities and cultural evolution. In Dunbar R, Barrett L, eds., *Oxford handbook of evolutionary psychology*, 555–570. New York: Oxford University Press.
Henshilwood CS, d'Errico F, Watts I. 2009. Engraved ochres from the Middle Stone Age levels at Blombos Cave, South Africa. *Journal of Human Evolution* 57(1):27–47.

Henshilwood CS, d'Errico F, Yates R, Jacobs Z, Tribolo C, Duller GA, Mercier N, Sealy JC, Valladas H, Watts I, et al. 2002. Emergence of modern human behavior: Middle Stone Age engravings from South Africa. *Science* 295(5558):1278–1280.

Henshilwood CS, Marean CW. 2003. The origin of modern human behavior. *Current Anthropology* 44(5):627–651.

Herbranson WT, Shimp CP. 2008. Artificial grammar learning in pigeons. *Learning & Behavior* 36(2):116–137.

Herrmann E, Call J, Hernández-Lloreda MV, Hare B, Tomasello M. 2007. Humans have evolved specialized skills of social cognition: The cultural intelligence hypothesis. *Science* 317(5843):1360–1366.

Herrnstein RJ. 1974. Formal properties of the matching law. *Journal of the Experimental Analysis of Behavior* 21(1):159.

Herzog HA. 2010. *Some we love, some we hate, some we eat: Why it's so hard to think straight about animals*. New York: HarperCollins.

Hesslow G. 2002. Conscious thought as simulation of behaviour and perception. *Trends in Cognitive Sciences* 6(6):242–247.

Heyes C. 1994. Social learning in animals: Categories and mechanisms. *Biological Reviews* 69(2):207–231.

Heyes C. 2001. Causes and consequences of imitation. *Trends in Cognitive Sciences* 5(6):253–261.

Heyes C. 2009. Evolution, development and intentional control of imitation. *Philosophical Transactions of the Royal Society B: Biological Sciences* 364(1528):2293–2298.

Heyes C. 2010. Where do mirror neurons come from? *Neuroscience & Biobehavioral Reviews* 34(4):575–583.

Heyes C. 2012a. Simple minds: A qualified defence of associative learning. *Philosophical Transactions of the Royal Society B* 367(1603):2695–2703.

Heyes C. 2012b. What's social about social learning? *Journal of Comparative Psychology* 126(2):193–202.

Heyes C. 2013. What can imitation do for cooperation? In Sterelny K, Joyce R, Calcott B, eds., *Cooperation and its evolution*, 313–332. Cambridge, MA: MIT Press.

Heyes C. 2016. Imitation: Not in our genes. *Current Biology* 26(10):R412–R414.

Heyes C. 2018. *Cognitive gadgets: The cultural evolution of thinking*. Cambridge, MA: Harvard University Press.

Heyes C, Galef BG Jr. 1996. *Social learning in animals: The roots of culture*. New York: Academic Press.

Heyes C, Jaldow E, Nokes T, Dawson G. 1994. Imitation in rats (*Rattus norvegicus*): The role of demonstrator action. *Behavioural Processes* 32(2):173–182.

Hilborn A, Pettorelli N, Orme CDL, Durant SM. 2012. Stalk and chase: How hunt stages affect hunting success in Serengeti cheetah. *Animal Behaviour* 84(3):701–706.

Hinde RA. 1970. *Animal behaviour: A synthesis of ethology and comparative psychology*. 2nd ed. Tokyo: McGraw-Hill Kogakusha.

Hinde RA, Stevenson-Hinde J, eds. 1973. *Constraints on learning*. London: Academic Press.

Hitchcock C. 2018. Causal models. In Zalta EN, ed., *The Stanford encyclopedia of philosophy*. Metaphysics Research Lab, Stanford University, fall 2018.

Hockett CF. 1960. The origin of speech. *Scientific American* 203:89–97.

Hodos W, Bonbright JC. 1972. The detection of visual intensity differences by pigeons. *Journal of the Experimental Analysis of Behavior* 18(3):471–479.

Hoff E. 2020. Lessons from the study of input effects on bilingual development. *International Journal of Bilingualism* 24(1):82–88.

Hoffmann DL, Standish CD, García-Diez M, Pettitt PB, Milton JA, Zilhão J, Alcolea-González JJ, Cantalejo-Duarte P, Collado H, de Balbín R, et al. 2018. U-Th dating of carbonate crusts reveals Neandertal origin of Iberian cave art. *Science* 359(6378):912–915.

Hogan JA. 2001. Development of behavior systems. In Blass E, ed., *Developmental psychobiology: Handbook of behavioral neurobiology*, vol. 13, 229–279. Dordrecht, Netherlands: Kluwer.

Hogan JA. 2017. *The study of behavior: Organization, methods, and principles*. Cambridge, UK: Cambridge University Press.

Hogan JA, Van Boxel F. 1993. Causal factors controlling dustbathing in Burmese red junglefowl: Some results and a model. *Animal Behaviour* 46:627–635.

Holland PC. 2008. Cognitive versus stimulus-response theories of learning. *Learning & Behavior* 36(3):227–241.

Hölldobler B, Wilson EO. 1990. *The Ants*. Cambridge, MA: Harvard University Press.

Hollis KL. 1997. Contemporary research on Pavlovian conditioning: A “new” functional analysis. *American Psychologist* 52(9):956.

Honey R, Hall G. 1989. Acquired equivalence and distinctiveness of cues. *Journal of Experimental Psychology: Animal Behavior Processes* 15(4):338.

Honig WK, James PHR, eds. 2016. *Animal memory*. New York: Academic Press.

Hoppitt W, Laland KN. 2013. *Social learning: An introduction to mechanisms, methods, and models*. Princeton, NJ: Princeton University Press.

Hoppitt W, Brown GR, Kendal R, Rendell L, Thornton A, Webster MM, Laland KN. 2008. Lessons from animal teaching. *Trends in Ecology and Evolution* 23(9):486–493.

Horner V, Whiten A. 2005. Causal knowledge and imitation/emulation switching in chimpanzees (*Pan troglodytes*) and children (*Homo sapiens*). *Animal Cognition* 8:164–181.

Houston AI, McNamara JM. 1999. *Models of adaptive behaviour*. Cambridge, UK: Cambridge University Press.

Houston AI, Trimmer PC, McNamara JM. 2021. Matching behaviours and rewards. *Trends in Cognitive Sciences* 25(5):403–415.

Howes MB. 2006. *Human memory: Structures and images*. Thousand Oaks, CA: Sage.

Hsiao HH. 1929. An experimental study of the rat's “insight” within a spatial complex. *University of California Publications in Psychology* 4(4):57–70.

Hubel DH. 1988. *Eye, brain, and vision*. New York: Scientific American Library.

Huber L, Rechberger S, Taborsky M. 2001. Social learning affects object exploration and manipulation in keas, *Nestor notabilis*. *Animal Behaviour* 62(5):945–954.

Hublin JJ, Ben-Ncer A, Bailey SE, Freidline SE, Neubauer S, Skinner MM, Bergmann I, Le Cabec A, Benazzi S, Harvati K, et al. 2017. New fossils from Jebel Irhoud, Morocco and the pan-African origin of *Homo sapiens*. *Nature* 546(7657):289.

Hudson JA. 2002. “Do you know what we're going to do this Summer?”: Mothers talk to preschool children about future events. *Journal of Cognition and Development* 3(1):49–71.

Hudson JA. 2006. The development of future time concepts through mother-child conversation. *Merrill-Palmer Quarterly* 52(1): 70–95.

Hudson JA, Fivush R. 1991. Planning in the preschool years: The emergence of plans from general event knowledge. *Cognitive Development* 6(4):393–415.

Hudson JA, Mayhew EM. 2011. Children's temporal judgments for autobiographical past and future events. *Cognitive Development* 26(4):331–342.

Hudson JA, Shapiro LR, Sosa BB. 1995. Planning in the real world: Preschool children's scripts and plans for familiar events. *Child Development* 66(4):984–998.

Hudson JA, Sosa BB, Shapiro LR. 2014. Scripts and plans: The development of preschool children's event knowledge and event planning. In Friedman S, Scholnick EK, eds., *The developmental psychology of planning: Why, how, and when do we plan?*, 89–114. Hove, East Sussex, UK: Psychology Press.

Hull CL. 1943. *Principles of behaviour*. New York: Appleton-Century-Crofts.

Hull CL. 1952. *A behavior system*. New Haven, CT: Yale University Press.

Humphrey NK. 1976. The social function of intellect. In Bateson PPG, Hinde RA, eds., *Growing points in ethology*, 303–317. Cambridge, UK: Cambridge University Press.

Iacoboni M, Dapretto M. 2006. The mirror neuron system and the consequences of its dysfunction. *Nature Reviews Neuroscience* 7(12):942–951.

Ifrah G. 1998. *The universal history of numbers from prehistory to the invention of the computer*. London: Harvill Press.

Immelmann K. 1972. The influence of early experience upon the development of social behaviour in estrildine finches. *Proceedings XVth Ornithological Congress, Den Haag 1970*, 316–338.

Inoue S, Matsuzawa T. 2007. Working memory of numerals in chimpanzees. *Current Biology* 17(23):R1004–R1005.

Inoue S, Matsuzawa T. 2009. Acquisition and memory of sequence order in young and adult chimpanzees (*Pan troglodytes*). *Animal Cognition* 12:S59–69.

Inoue-Nakamura N, Matsuzawa T. 1997. Development of stone tool use by wild chimpanzees (*Pan troglodytes*). *Journal of Comparative Psychology* 11(2):159–173.

Itan Y, Powell A, Beaumont MA, Burger J, Thomas MG. 2009. The origins of lactase persistence in Europe. *PLoS Computational Biology* 5(8):e1000491.

Jablonka E, Lamb MJ. 2005. *Evolution in four dimensions*. Cambridge, MA: MIT Press.

Jackson A. 2002. The world of blind mathematicians. *Notices of the American Mathematical Society* 49(10):1246–1251.

Jackson M. 2012. Machiavellian intelligence hypothesis. In Seel NM, ed., *Encyclopedia of the sciences of learning*, 2081–2082. Boston: Springer.

Janson CH. 2014. Death of the (traveling) salesman: Primates do not show clear evidence of multi-step route planning. *American Journal of Primatology* 76(5):410–420.

Janson CH. 2016. Capuchins, space, time and memory: An experimental test of what-where-when memory in wild monkeys. *Proceedings of the Royal Society B: Biological Sciences* 283(1840):20161432.

Janson CH, Byrne R. 2007. What wild primates know about resources: Opening up the black box. *Animal Cognition* 10(3):357–367.

Jelbert SA, Taylor AH, Cheke LG, Clayton NS, Gray RD. 2014. Using the Aesop's fable paradigm to investigate causal understanding of water displacement by New Caledonian crows. *PLoS One* 9(3):e92895.

Jenkins H, Moore BR. 1973. The form of the auto-shaped response with food or water reinforcers. *Journal of the Experimental Analysis of Behavior* 20(2):163–181.

Jensen R. 2006. Behaviorism, latent learning, and cognitive maps: Needed revisions in introductory psychology textbooks. *Behavior Analyst* 29(2):187–209.

Johnson-Laird PN. 2006. *How we reason.* New York: Oxford University Press.

Jolly CJ. 1970. The seed-eaters: A new model of hominid differentiation based on a baboon analogy. *Man* 5(1):5–26.

Jolly CJ, Webb JK, Phillips BL. 2018. The perils of paradise: An endangered species conserved on an island loses antipredator behaviours within 13 generations. *Biology Letters* 14(6):20180222.

Jonsson M, Ghirlanda S, Lind J, Vinken V, Enquist M. 2021. Learning simulator: A simulation software for animal and human learning. *Journal of Open Source Software* 6(58):2891.

Jordan PD. 2014. *Technology as human social tradition.* Berkeley: University of California Press.

Jusczyk PW. 1997. *The discovery of spoken language.* Cambridge, MA: MIT Press.

Kaas JH. 2016. *Evolution of nervous systems.* New York: Academic Press.

Kahneman D. 2011. *Thinking, fast and slow.* New York: Farrar, Strauss and Giroux.

Kaller CP, Rahm B, Spreer J, Mader I, Unterrainer JM. 2008. Thinking around the corner: The development of planning abilities. *Brain and Cognition* 67(3):360–370.

Kamil AC, Balda RP. 1985. Cache recovery and spatial memory in Clark's nutcrackers (*Nucifraga columbiana*). *Journal of Experimental Psychology: Animal Behavior Processes* 11(1):95.

Kamil AC, Roitblat HL. 1985. The ecology of foraging behavior: Implications for animal learning and memory. *Annual Review of Psychology* 36(1):141–169.

Kamin LJ. 1969. Predictability, surprise, attention, and conditioning. In Campbell BA, Church MR, eds., *Punishment and aversive behavior*, 279–296. New York: Appleton-Century-Crofts.

Kaplan H, Bock J. 2001. Fertility theory: The embodied capital theory of life history evolution. In Hoem JM, ed., *International encyclopedia of the social and behavioral sciences*, 5561–5568. New York: Elsevier.

Kaplan H, Hill K, Lancaster J, Hurtado AM. 2000. A theory of human life history evolution: Diet, intelligence, and longevity. *Evolutionary Anthropology: Issues, News, and Reviews* 9(4):156–185.

Kaplan H, Lancaster J, Robson A. 2003. Embodied capital and the evolutionary economics of the human life span. *Population and Development Review* 29:152–182.

Kaplan H, Robson A. 2002. The emergence of humans: The coevolution of intelligence and longevity with intergenerational transfers. *Proceedings of the National Academy of Sciences* 99(15):10221–10226.

Karkanas P, Shahack-Gross R, Ayalon A, Bar-Matthews M, Barkai R, Frumkin A, Gopher A, Stiner MC. 2007. Evidence for habitual use of fire at the end of the Lower Paleolithic: Site-formation processes at Qesem Cave, Israel. *Journal of Human Evolution* 53(2):197–212.

Karmiloff Smith A. 1992. *Beyond modularity: A developmental perspective on cognitive science.* Cambridge, MA: MIT Press.

Keefner A. 2016. Corvids infer the mental states of conspecifics. *Biology & Philosophy* 31(2):267–281.

Kelleher RT, Gollub LR. 1962. A review of positive conditioned reinforcement. *Journal of the Experimental Analysis of Behavior* 5(4):543–597.

Kellogg W, Kellogg L. 1933. *The ape and the child: A study of early environmental influence upon early behavior*. New York: McGraw-Hill.

Kendler HH. 1947. An investigation of latent learning in a T-maze. *Journal of Comparative and Physiological Psychology* 40(4):265.

Kendrick DF, Rilling ME, Denny MR, eds. 1986. *Theories of animal memory*. Hillsdale, NJ: Lawrence Erlbaum.

Kenward B, Rutz C, Weir AA, Kacelink A. 2006. Development of tool use in New Caledonian crows: Inherited action patterns and social influences. *Animal Behavior* 72:1329–1343.

Kenward B, Weir AA, Rutz C, Kacelnik A. 2005. Tool manufacture by naive juvenile crows. *Nature* 433(7022):121.

Keogh E, Mueen A. 2010. Curse of dimensionality. In Sammut C, Webb GI, eds., *Encyclopedia of machine learning*, 257–258. New York: Springer.

King PM, Kitchener KS. 1994. *Developing reflective judgment: Understanding and promoting intellectual growth and critical thinking in adolescents and adults*. San Francisco: Jossey-Bass.

Kis A, Huber L, Wilkinson A. 2014. Social learning by imitation in a reptile (*Pogona vitticeps*). *Animal Cognition* 18(1):325–331.

Klein RG. 1995. Anatomy, behavior, and modern human origins. *Journal of World Prehistory* 9(2):167–198.

Klein RG. 2001. Southern Africa and modern human origins. *Journal of Anthropological Research* 57(1):1–16.

Klein RG. 2002. *The dawn of human culture*. New York: Wiley & Sons.

Klein RG. 2008. Out of Africa and the evolution of human behavior. *Evolutionary Anthropology: Issues, News, and Reviews* 17(6):267–281.

Klein SB, Cosmides L, Gangi CE, Jackson B, Tooby J, Costabile KA. 2009. Evolution and episodic memory: An analysis and demonstration of a social function of episodic recollection. *Social Cognition* 27(2):283–319.

Klette R. 2014. *Concise computer vision*. Berlin: Springer.

Kobayashi Y, Aoki K. 2012. Innovativeness, population size and cumulative cultural evolution. *Theoretical Population Biology* 82(1):38–47.

Koenen C, Pusch R, Bröker F, Thiele S, Güntürkün O. 2016. Categories in the pigeon brain: A reverse engineering approach. *Journal of the Experimental Analysis of Behavior* 105(1):111–122.

Köhler W. 1924. *The mentality of apes*. New York: Harcourt Brace.

Kolodny O, Creanza N, Feldman MW. 2015. Evolution in leaps: The punctuated accumulation and loss of cultural innovations. *Proceedings of the National Academy of Sciences* 112(49):E6762–E6769.

Konner M. 2010. *The evolution of childhood: Relationships, emotion, mind*. Cambridge, MA: Harvard University Press.

Koops K, Furuichi T, Hashimoto C. 2015. Chimpanzees and bonobos differ in intrinsic motivation for tool use. *Scientific Reports* 5:11356.

Kovas Y, Garon-Carrier G, Boivin M, Petrill SA, Plomin R, Malykh SB, Spinath F, Murayama K, Ando J, Bogdanova OY, et al. 2015. Why children differ in motivation to learn: Insights from over 13,000 twins from 6 countries. *Personality and Individual Differences* 80:51–63.

Krause J, Lalueza-Fox C, Orlando L, Enard W, Green RE, Burbano HA, Hublin JJ, Hänni C, Fortea J, De La Rasilla M, et al. 2007. The derived FOXP2 variant of modern humans was shared with Neandertals. *Current Biology* 17(21):1908–1912.

Krebs JR, Davies NB, eds. 1978. *Behavioural ecology: An evolutionary approach.* London: Blackwell Scientific.

Krebs JR, Erichsen JT, Webber MI, Charnov EL. 1977. Optimal prey selection in the great tit (*Parus major*). *Animal Behaviour* 25:30–38.

Kristo G, Janssen SM, Murre JM. 2009. Retention of autobiographical memories: An internet-based diary study. *Memory* 17(8):816–829.

Kroeber AL. 1917. The Superorganic. *American Anthropologist* 19:163–213.

Kuba MJ, Byrne RA, Meisel DV, Mather JA. 2006. When do octopuses play? Effects of repeated testing, object type, age, and food deprivation on object play in *Octopus vulgaris*. *Journal of Comparative Psychology* 120(3):184.

Kuhn D, Pearsall S. 2000. Developmental origins of scientific thinking. *Journal of Cognition and Development* 1(1):113–129.

Kuhn TS. 1996. *The structure of scientific revolutions.* Chicago: University of Chicago Press.

Kullberg C, Lind J. 2002. An experimental study of predator recognition in great tit fledglings. *Ethology* 108:429–441.

Kutlu MG, Schmajuk NA. 2012. Classical conditioning mechanisms can differentiate between seeing and doing in rats. *Journal of Experimental Psychology: Animal Behavior Processes* 38(1):84.

Lai M. 2015. Giraffe: Using deep reinforcement learning to play chess. Master's thesis, Imperial College London.

Laland KN. 2004. Social learning strategies. *Learning and Behavior* 32:4–14.

Laland KN. 2008. Exploring gene-culture interactions: Insights from handedness, sexual selection, and niche-construction case studies. *Philosophical Transactions of the Royal Society B* 363:3577–3589.

Laland KN, Brown GR. 2011. *Sense and nonsense: Evolutionary perspectives on human behaviour.* New York: Oxford University Press.

Laland KN, Galef BG. Jr. 2009. *The question of animal culture.* Cambridge, MA: Harvard University Press.

Laland KN, Odling Smee J, Feldman MW. 2001. Cultural niche construction and human evolution. *Journal of Evolutionary Biology* 14:22–33.

Lancaster JB, Lancaster CS. 1983. Parental investment: The hominid adaptation. In Ortner DJ, ed., *How humans adapt: A biocultural odyssey*, 33–56. Washington, DC: Smithsonian Institution Press.

Langley MC, Clarkson C, Ulm S. 2008. Behavioural complexity in Eurasian Neanderthal populations: A chronological examination of the archaeological evidence. *Cambridge Archaeological Journal* 18(03):289–307.

Lapiedra O, Schoener TW, Leal M, Losos JB, Kolbe JJ. 2018. Predator-driven natural selection on risk-taking behavior in anole lizards. *Science* 360(6392):1017–1020.

Lara C, González JM, Hudson R. 2009. Observational learning in the white-eared hummingbird (*Hylocharis leucotis*): Experimental evidence. *Ethology* 115(9):872–878.

Lashley KS. 1951. The problem of serial order in behavior. In Jeffress LA, ed., *Cerebral mechanisms in behavior: The Hixon Symposium*, 112–146. New York: Wiley.

Lefebvre L, Templeton J, Brown K, Koelle M. 1997. Carib grackles imitate conspecific and Zenaida dove tutors. *Behaviour* 134(13–14):1003–1017.

Lehman HC. 1947. The exponential increase in man's cultural output. *Social Forces* 25(3):281–290.

Lehrman DS. 1953. A critique of Konrad Lorenz's theory of instinctive behavior. *Quarterly Review of Biology* 22(4):337–363.

Leigh EG Jr. 2010. The evolution of mutualism. *Journal of Evolutionary Biology* 23(12):2507–2528.

Lemaire P, Arnaud L. 2008. Young and older adults' strategies in complex arithmetic. *American Journal of Psychology* 121(1):1–16.

Levinson SC, Enfield NJ. 2020. New York: *Roots of human sociality: Culture, cognition and interaction*. New York: Routledge.

Levitsky D, Collier G. 1968. Schedule-induced wheel running. *Physiology & Behavior* 3(4):571–573.

Lewis HM, Laland KN. 2012. Transmission fidelity is the key to the build-up of cumulative culture. *Philosophical Transactions of the Royal Society B* 367(1599):2171–2180.

Lewis K. 2003. Measuring transactive memory systems in the field: Scale development and validation. *Journal of Applied Psychology* 88(4):587.

Lewis M. 1983. *Origins of intelligence: Infancy and early childhood*. New York: Springer.

Lew-Levy S, Reckin R, Lavi N, Cristóbal-Azkarate J, Ellis-Davies K. 2017. How do hunter-gatherer children learn subsistence skills? *Human Nature* 28(4):367–394.

Lieberman DA. 2011. *Human learning and memory*. Cambridge, UK: Cambridge University Press.

Limongelli L, Boysen ST, Visalberghi E. 1995. Comprehension of cause-effect relations in a tool-using task by chimpanzees (*Pan troglodytes*). *Journal of Comparative Psychology* 109(1):18–26.

Lind J. 2018. What can associative learning do for planning? *Royal Society Open Science* 5(11):180778.

Lind J, Enquist M, Ghirlanda S. 2015. Animal memory: A review of delayed matching-to-sample data. *Behavioural Processes* 117:52–58.

Lind J, Ghirlanda S, Enquist M. 2009. Insight learning or shaping? *Proceedings of the National Academy of Sciences* 106(28):E76.

Lind J, Ghirlanda S, Enquist M. 2019. Social learning through associative processes: A computational theory. *Royal Society Open Science* 6:181777.

Lind J, Ghirlanda S, Enquist M. 2021. Evolution of memory systems in animals. In Krause M, Hollis KL, Papini MR, eds., *Evolution of learning and memory mechanisms*. 339–358. Cambridge, UK: Cambridge University Press.

Lind J, Lindenfors P, Ghirlanda S, Lidén K, Enquist M. 2013. Dating human cultural capacity using phylogenetic principles. *Scientific Reports* 3:1785.

Lind J, Lönnberg S, Persson T, Enquist M. 2017. Time does not help orangutans *Pongo abelii* solve physical problems. *Frontiers in Psychology* 8:161.

LoBue V, Rakison DH. 2013. What we fear most: A developmental advantage for threat-relevant stimuli. *Developmental Review* 33(4):285–303.

Logue AW. 1988. Research on self-control: An integrating framework. *Behavioral and Brain Sciences* 11(4):665–679.

LoLordo VM. 1979. Selective associations. In Dickinson A, Boakes RA, eds., *Mechanisms of learning and motivation: A memorial volume to Jerzy Konorski*, 367–398. Hillsdale, NJ: Lawrence Erlbaum.

Lonsdorf EV. 2005. Sex differences in the development of termite-fishing skills in the wild chimpanzees, *Pan troglodytes schweinfurthii*, of Gombe National Park, Tanzania. *Animal Behaviour* 70(3):673–683.

Lonsdorf EV, Bonnie KE. 2010. Opportunities and constraints when studying social learning: Developmental approaches and social factors. *Learning & Behavior* 38(3):195–205.

Lorenz K. 1941. Vergleichende bewegungsstudien bei anatiden. *Journal für Ornithologie* 89:194–294.

Lorenz K. 1973. *behind the mirror: A search for a natural history of human knowledge*. New York: Harcourt Brace Jovanovich.

Loucks J, Mutschler C, Meltzoff AN. 2017. Children's representation and imitation of events: How goal organization influences 3-year-old children's memory for action sequences. *Cognitive Science* 41(7):1904–1933.

Loucks J, Price HL. 2019. Memory for temporal order in action is slow developing, sensitive to deviant input, and supported by foundational cognitive processes. *Developmental Psychology* 55(2):263.

Lovejoy C. 1981. The origin of man. *Science* 211(4480):341–350.

Lucas AJ, Kings M, Whittle D, Davey E, Happé F, Caldwell CA, Thornton A. 2020. The value of teaching increases with tool complexity in cumulative cultural evolution. *Proceedings of the Royal Society B* 287(1939):20201885.

Luce MR, Callanan MA, Smilovic S. 2013. Links between parents' epistemological stance and children's evidence talk. *Developmental Psychology* 49(3):454.

Luciano M, Wright M, Smith G, Geffen G, Geffen L, Martin N. 2001. Genetic covariance among measures of information processing speed, working memory, and IQ. *Behavior Genetics* 31(6):581–592.

Ludvig EA, Mirian MS, Kehoe EJ, Sutton RS. 2017. Associative learning from replayed experience. *bioRxiv*. https://doi.org/10.1101/100800.

Luenberger DG. 1979. *Introduction to dynamic systems: Theory, models and applications*. New York: Wiley.

Lumsden CJ, Wilson EO. 1981. *Genes, mind, and culture: The coevolutionary process*. Cambridge, MA: Harvard University Press.

MacCorquodale K, Meehl PE, Tolman EC. 1954. In Estes WK, Koch S, MacCorquodale K, Meehl PE, Mueller CG, Schoenfeld WN, Verplanck WS, eds., *Modern learning theory*, 177–266. New York: Appleton-Century-Crofts.

MacDonald SE. 1993. Delayed matching-to-successive-samples in pigeons: Short-term memory for item and order information. *Animal Learning & Behavior* 21(1):59–67.

Mackintosh NJ. 1969. Further analysis of the overtraining reversal effect. *Journal of Comparative and Physiological Psychology* 67(2, pt.2):1–18.

Mackintosh NJ. 1983a. General principles of learning. In Halliday TR, Slater PJ, eds., *Animal behaviour: Genes, development and learning*, vol. 3, 178–212. New York: W. H. Freeman.

Mackintosh NJ. 1974. *The psychology of animal learning*. New York: Academic Press.

Mackintosh NJ. 1983b. *Conditioning and associative learning*. New York: Oxford University Press.

Mackintosh NJ. 2011. *IQ and human intelligence*. New York: Oxford University Press.

MacLean EL. 2016. Unraveling the evolution of uniquely human cognition. *Proceedings of the National Academy of Sciences* 113(23):6348–6354.

MacLean EL, Hare B, Nunn CL, Addessi E, Amici F, Anderson RC, Aureli F, Baker JM, Bania AE, Barnard AM, et al. 2014. The evolution of self-control. *Proceedings of the National Academy of Sciences* 111(20):E2140–E2148.

Macphail EM. 1982. *Brain and intelligence in vertebrates*. Oxford, UK: Clarendon Press.

Macphail EM, Barlow H. 1985. Vertebrate intelligence: The null hypothesis [and discussion]. *Philosophical Transactions of the Royal Society B* 308(1135):37–51.

Macphail EM, Bolhuis JJ. 2001. The evolution of intelligence: Adaptive specializations versus general process. *Biological Reviews* 76(3):341–364.

Mahr JB, Csibra G. 2018. Why do we remember? The communicative function of episodic memory. *Behavioral and Brain Sciences* 41.

Marler P, Tamura M. 1964. Culturally transmitted patterns of vocal behaviour in sparrows. *Science* 146:1483–1486.

Marsland TA. 1990. A short history of computer chess. In Marsland TA, Schaeffer J, eds., *Computers, chess, and cognition*, 3–7. New York: Springer.

Maslin MA, Shultz S, Trauth MH. 2015. A synthesis of the theories and concepts of early human evolution. *Philosophical Transactions of the Royal Society B* 370(1663):20140064.

Mather JA, Anderson RC. 1999. Exploration, play and habituation in octopuses (*Octopus dofleini*). *Journal of Comparative Psychology* 113(3):333.

Mauk MD, Donegan NH. 1997. A model of Pavlovian eyelid conditioning based on the synaptic organization of the cerebellum. *Learning & Memory* 4:130–158.

Maynard Smith J. 1982. *Evolution and the theory of games*. Cambridge, UK: Cambridge University Press.

Maynard Smith J, Szathmáry E. 1995. *The major transitions in evolution*. New York: W. H. Freeman.

McBrearty S, Brooks AS. 2000. The revolution that wasn't: A new interpretation of the origin of modern human behavior. *Journal of Human Evolution* 39(5):453–563.

McCormack T, Atance CM. 2011. Planning in young children: A review and synthesis. *Developmental Review* 31(1):1–31.

McElreath R, Henrich J. 2007. Modeling cultural evolution. In Dunbar R, Barrett L, eds., *Oxford handbook of evolutionary psychology*, 571–586. New York: Oxford University Press.

McFarland D. 1971. *Feedback mechanisms in animal behaviour*. New York: Academic Press.

McFarland D. 1985. *Animal behaviour*. Boston: Pitman.

McFarland D, Bösser T. 1993. *Intelligent behavior in animals and robots*. Cambridge, MA: MIT Press.

McFarland D, Houston A. 1981. *Quantitative ethology: The state space approach*. Boston: Pitman.

McGreevy P, Boakes R. 2011. *Carrots and sticks: Principles of animal training*. Sydney: Darlington Press.

McLaren IPL, Mackintosh NJ. 2000. An elemental model of associative learning: I. Latent inhibition and perceptual learning. *Animal Learning & Behavior* 28(3):211–246.

McLaren IPL, Mackintosh NJ. 2002. Associative learning and elemental representation: II. Generalization and discrimination. *Animal Learning & Behavior* 30:177–200.

McNally L, Jackson AL. 2013. Cooperation creates selection for tactical deception. *Proceedings of the Royal Society B* 280(1762):20130699.

Meadows S. 1993. *The child as thinker*. London: Routledge.

Melis AP, Semmann D. 2010. How is human cooperation different? *Philosophical Transactions of the Royal Society B* 365(1553):2663–2674.

Meltzoff AN, Murray L, Simpson E, Heimann M, Nagy E, Nadel J, Pedersen EJ, Brooks R, Messinger DS, Pascalis LD, et al. 2018. Re-examination of Oostenbroek et al. (2016): Evidence for neonatal imitation of tongue protrusion. *Developmental Science* 21(4):e12609.

Menzel EW. 1973a. Chimpanzee spatial memory organization. *Science* 182(4115):943–945.

Menzel EW, ed. 1973b. *Precultural primate behavior*. Basel, Switzerland: Karger.

Mercado E III, Perazio CE. 2021. Similarities in composition and transformations of songs by humpback whales (*Megaptera novaeangliae*) over time and space. *Journal of Comparative Psychology* 135(1):28.

Mesoudi A. 2011. *Cultural evolution: How darwininan evolutionary theory can explain human culture and synthesize the social Sciences*. Chicago: University of Chicago Press.

Mesoudi A. 2016. Cultural evolution: A review of theory, findings and controversies. *Evolutionary Biology* 43(4):481–497.

Metzgar LH. 1967. An experimental comparison of screech owl predation on resident and transient white-footed mice (*Peromyscus leucopus*). *Journal of Mammalogy* 48(3):387–391.

Michaelian K. 2016. *Mental time travel: Episodic memory and our knowledge of the personal past*. Cambridge, MA: MIT Press.

Miller GA. 1956. The magical number seven, plus or minus two: Some limits on our capacity for processing information. *Psychological Review* 63(2):81.

Miller GA. 2003. The cognitive revolution: A historical perspective. *Trends in Cognitive Sciences* 7(3):141–144.

Miller GA, Eugene G, Pribram KH. 1960. *Plans and the structure of behaviour*. New York: Holt, Rinehart and Winston.

Miller N. 2018. Social learning and associative processes: A synthesis. *Journal of Experimental Psychology: Animal Learning and Cognition* 44(2):105.

Miller NE. 1935. A reply to "Sign-gestalt or conditioned reflex?" *Psychological Review* 42:280–292.

Mineka S. 1984. Observational conditoning of snake fear in rhesus monkeys. *Journal of Abnormal Psychology* 93:355–372.

Mineka S, Cook M. 1988. Social learning and the acquisition of snake fear in monkeys. In Zentall TR, Galef BG Jr., eds., *Social learning: Psychological and biological perspectives*, 51–73. Mahwah, NJ: Lawrence Erlbaum.

Mithen S. 1996. *The prehistory of the mind: The cognitive origins of art and science*. London: Thames & Hudson.

Moore BR. 1973. The role of directed Pavlovian reactions in simple instrumental learning in the pigeon. In Hinde RA, Stevenson-Hinde J, eds., *Constraints on learning*, 159–188. London: Academic Press.

Moore C, Dunham PJ. 1995. *Joint attention: Its origins and role in development*. Hove, East Sussex, UK: Psychology Press.

Moore R. 2021. The cultural evolution of mind-modelling. *Synthese* 199:1751–1776.

Morales M, Mundy P, Delgado CE, Yale M, Messinger D, Neal R, Schwartz HK. 2000. Responding to joint attention across the 6-through 24-month age period and early language acquisition. *Journal of Applied Developmental Psychology* 21(3):283–298.

Morgan JL. 1989. Learnability considerations and the nature of trigger experiences in language acquisition. *Behavioral and Brain Sciences* 12(02):352–353.

Morin E, Laroulandie V. 2012. Presumed symbolic use of diurnal raptors by Neanderthals. *PLoS One* 7(3):e32856.

Motes-Rodrigo A, Mundry R, Call J, Tennie C. 2021. Evaluating the influence of action- and subject-specific factors on chimpanzee action copying. *Royal Society Open Science* 8(2):200228.

Mueller T, O'Hara RB, Converse SJ, Urbanek RP, Fagan WF. 2013. Social learning of migratory performance. *Science* 341(6149):999–1002.

Mulcahy NJ, Call J. 2006. Apes save tools for future use. *Science* 312(5776):1038–1040.

Munakata Y. 2006. Information processing approaches to development. In Kuhn D, Siegler RS, Damon W, Lerner RM, eds., *Handbook of child psychology: Cognition, perception, and language*, 426–463. New York: Wiley.

Munro T. 1963. *Evolution in the arts and other theories of culture history*. Cleveland, OH: Cleveland Museum of Art.

Musgrave JH. 1971. How dextrous was Neanderthal man? *Nature* 233:538–541.

Muthukrishna M, Doebeli M, Chudek M, Henrich J. 2018. The cultural brain hypothesis: How culture drives brain expansion, sociality, and life history. *PLoS Computational Biology* 14(11):e1006504.

Myowa Yamakoshi M, Matsuzawa T. 1999. Factors influencing imitation of manipulatory actions in chimpanzees. *Journal of Comparative Psychology* 113:128–136.

Nakajima S, Sato M. 1993. Removal of an obstacle: Problem-solving behavior in pigeons. *Journal of the Experimental Analysis of Behavior* 59(1):131–145.

Neadle D, Allritz M, Tennie C. 2017. Food cleaning in gorillas: Social learning is a possibility but not a necessity. *PLoS One* 12(12):e0188866.

Nelson K. 1998. *Language in cognitive development: The emergence of the mediated mind*. Cambridge, UK: Cambridge University Press.

Nelson K. 2009. *Young minds in social worlds: Experience, meaning, and memory*. Cambridge, MA: Harvard University Press.

Nelson K, Fivush R. 2004. The emergence of autobiographical memory: A social cultural developmental theory. *Psychological Review* 111(2):486.

Nelson K, Gruendel J. 1981. Generalized event representations: Basic building blocks of cognitive development. *Advances in Developmental Psychology* 1:131–158.

Newell A, Rosenbloom PS. 1981. Mechanisms of skill acquisition and the law of practice. *Cognitive Skills and Their Acquisition* 1:1–55.

Newell A, Shaw JC, Simon HA. 1959. Report on a general problem-solving program. *Proceedings of the International Conference on Information Processing*, 256–264.

Newell A, Simon HA 1972. *Human problem solving*. Englewood Cliffs, NJ: Prentice Hall.

Niewoehner WA, Bergstrom A, Eichele D, Zuroff M, Clark JT. 2003. Digital analysis: Manual dexterity in Neanderthals. *Nature* 422(6930):395.

Nilsson NJ. 2009. *The quest for artificial intelligence*. Cambridge, UK: Cambridge University Press.

Nisbett RE, Aronson J, Blair C, Dickens W, Flynn J, Halpern DF, Turkheimer E. 2012. Intelligence: New findings and theoretical developments. *American Psychologist* 67(2):130.

Nishida T, Hiraiwa M. 1982. Natural history of a tool-using behavior by wild chimpanzees in feeding upon wood-boring ants. *Journal of Human Evolution* 11(1):73–99.

Nishihara T, Suzuki S, Kuroda S. 1995. Tool-set for termite-fishing by chimpanzees in the Ndoki Forest, Congo. *Behaviour* 132(3–4):219–235.

Nowak M, Highfield R. 2011. *Supercooperators: Altruism, evolution, and why we need each other to succeed*. Simon & Schuster.

Nowell A. 2010. Defining behavioral modernity in the context of Neandertal and anatomically modern human populations. *Annual Review of Anthropology* 39:437–452.

Nunn J, Burgess G. 2009. *Understanding chess endgames*. London: Gambit.

Odling Smee F, Laland K, Feldman M. 2003. *Niche construction: The neglected process in evolution*. Princeton, NJ: Princeton University Press.

Ogburn WF. 1950. *Social change*. New York: Viking.

Ohlsson S. 2012. The problems with problem solving: Reflections on the rise, current status, and possible future of a cognitive research paradigm. *Journal of Problem Solving* 5(1):7.

Olson DJ, Kamil AC, Balda RP, Nims PJ. 1995. Performance of four-seed caching corvid species in operant tests of nonspatial and spatial memory. *Journal of Comparative Psychology* 109(2):173.

Oostenbroek J, Suddendorf T, Nielsen M, Redshaw J, Kennedy-Costantini S, Davis J, Clark S, Slaughter V. 2016. Comprehensive longitudinal study challenges the existence of neonatal imitation in humans. *Current Biology* 26(10):1334–1338.

Ormerod BK, Beninger RJ. 2002. Water maze versus radial maze: Differential performance of rats in a spatial delayed match-to-position task and response to scopolamine. *Behavioural Brain Research* 128(2):139–152.

Ostrom E. 2010. Beyond markets and states: Polycentric governance of complex economic systems. *Transnational Corporations Review* 2(2):1–12.

Ostrom E. 2015. *Governing the commons*. Cambridge, UK: Cambridge University Press.

Pahl M, Zhu H, Pix W, Tautz J, Zhang S. 2007. Circadian timed episodic-like memory—a bee knows what to do when, and also where. *Journal of Experimental Biology* 210(20):3559–3567.

Panksepp J. 1998. *Affective neuroscience: The foundations of human and animal emotions*. New York: Oxford University Press.

Panksepp J. 2005. Affective consciousness: Core emotional feelings in animals and humans. *Consciousness and Cognition* 14(1):30–80.

Papa A, Cristea M, McGuigan N, Tamariz M. 2020. The role of language in cultural evolution. *PsyArXiv*. https://doi.org/10.31234/osf.io/jqemk.

Parker CE. 1974. Behavioral diversity in ten species of nonhuman primates. *Journal of Comparative and Physiological Psychology* 87(5):930.

Parker CE. 1978. Opportunism and the rise of intelligence. *Journal of Human Evolution* 7(7):597–608.

Parker ST, Poti P. 1990. The role of innate motor patterns in ontogenetic and experimental development of intelligent use of sticks in cebus monkeys. In Parker ST, Gibson KR, eds., *"Language" and intelligence in monkeys and apes*, 219–243. Cambridge, UK: Cambridge University Press.

Parkinson JA, Roberts AC, Everitt BJ, Di Ciano P. 2005. Acquisition of instrumental conditioned reinforcement is resistant to the devaluation of the unconditioned stimulus. *Quarterly Journal of Experimental Psychology* B58(1b):19–30.

Pearce JM. 2008. *Animal learning and cognition*. 3rd ed. Hove, East Sussex, UK: Psychology Press.

Pellis SM. 1991. How motivationally distinct is play? A preliminary case study. *Animal Behaviour* 42(5):851–853.

Penn DC, Holyoak KJ, Povinelli DJ. 2008. Darwin's mistake: Explaining the discontinuity between human and nonhuman minds. *Behavioral and Brain Sciences* 31(02): 109–130.

Perkins DN, Salomon G. 1989. Are cognitive skills context-bound? *Educational Researcher* 18(1):16–25.

Perry SE, Manson JH. 2003. Traditions in monkeys. *Evolutionary Anthropology* 12:71–81.

Persson T, Sauciuc GA, Madsen EA. 2018. Spontaneous cross-species imitation in interactions between chimpanzees and zoo visitors. *Primates* 59(1):19–29.

Pezdek K, Miceli L. 1982. Life-span differences in memory integration as a function of processing time. *Developmental Psychology* 18(3):485.

Piaget J. 1977. *The origin of intelligence in the child*. Penguin Books.

Pica P, Lemer C, Izard V, Dehaene S. 2004. Exact and approximate arithmetic in an Amazonian indigene group. *Science* 306(5695):499–503.

Pierce WD, Cheney CD. 2018. *Behavior analysis and learning*. Hove, East Sussex, UK: Psychology Press.

Pierrel R, Sherman JG. 1963. Barnabus, the rat with college training. *Brown Alumni Monthly* 63(5):9–12. https://archive.org/details/brownalumnimonth635brow/mode/2up.

Pilley JW, Reid AK. 2011. Border collie comprehends object names as verbal referents. *Behavioural Processes* 86(2):184–195.

Pinker S. 1994. *The language instinct*. London: Penguin.

Pinker S. 2002. *The blank slate: The modern denial of human nature*. London: Penguin.

Pladevall J, Mendes N, Riba D, Llorente M, Amici F. 2020. No evidence of what-where-when memory in great apes (*Pan troglodytes*, *Pan paniscus*, *Pongo abelii*, and *Gorilla gorilla*). *Journal of Comparative Psychology* 134(2):252–261.

Plomin R, Petrill SA. 1997. Genetics and intelligence: What's new? *Intelligence* 24(1):53–77.

Poirier FE, Smith EO. 1974. Socializing functions of primate play. *American Zoologist* 14(1):275–287.

Polderman TJ, Stins JF, Posthuma D, Gosso MF, Verhulst FC, Boomsma DI. 2006. The phenotypic and genotypic relation between working memory speed and capacity. *Intelligence* 34(6):549–560.

Polya G. 2004. *How to solve it: A new aspect of mathematical method.* Princeton, NJ: Princeton University Press.

Potts R. 1998. Environmental hypotheses of hominin evolution. *American Journal of Physical Anthropology* 107(27):93–136.

Potts R. 2004. Paleoenvironmental basis of cognitive evolution in great apes. *American Journal of Primatology* 62(3):209–228.

Poundstone W. 1992. *Prisoner's dilemma: John von Neumann, game theory and the puzzle of the bomb.* New York: Doubleday.

Povinelli DJ. 2020. Can comparative psychology crack its toughest nut? *Animal Behavior and Cognition* 7(4):589–652.

Povinelli DJ, Preuss TM. 1995. Theory of mind: Evolutionary history of a cognitive specialization. *Trends in Neurosciences* 18(9):418–424.

Powell A, Shennan S, Thomas MG. 2009. Late Pleistocene demography and the appearance of modern human behavior. *Science* 324:1298–1301.

Premack D. 1983. Animal cognition. *Annual Review of Psychology* 34(1):351–362.

Premack D, Premack A. 1996. Why animals lack pedagogy and some cultures have more of it than others. In Olson DR, Torrance N, eds., *The handbook of human development and education*, 302–344. London: Blackwell.

Pressley M, Woloshyn V. 1995. *Cognitive strategy instruction.* Northampton, MA: Brookline.

Prinz J. 2012. *Beyond human nature.* New York: W. W. Norton.

Prüfer K, Racimo F, Patterson N, Jay F, Sankararaman S, Sawyer S, Heinze A, Renaud G, Sudmant PH, De Filippo C, et al. 2014. The complete genome sequence of a Neanderthal from the Altai Mountains. *Nature* 505(7481):43.

Quiroga RQ. 2020. No pattern separation in the human hippocampus. *Trends in Cognitive Sciences* 24(12):994–1007.

Raby CR, Alexis DM, Dickinson A, Clayton NS. 2007. Planning for the future by western scrub-jays. *Nature* 445(7130):919–921.

Raby CR, Clayton NS. 2009. Prospective cognition in animals. *Behavioural Processes* 80(3):314–324.

Radziszewska B, Rogoff B. 1988. Influence of adult and peer collaborators on children's planning skills. *Developmental Psychology* 24(6):840.

Radziszewska B, Rogoff B. 1991. Children's guided participation in planning imaginary errands with skilled adult or peer partners. *Developmental Psychology* 27(3):381.

Ramus F. 2017. General intelligence is an emerging property, not an evolutionary puzzle. *Behavioral and Brain Sciences* 40:E217.

Rashotte M. 1979. Reward training: Latent learning. In Bitterman M, Lolordo V, Overmier J, Rahsotte M, eds., *Animal learning: Survey and analysis*, 167–193. New York: Springer.

Read DW, Manrique HM, Walker MJ. 2022. On the working memory of humans and great apes: Strikingly similar or remarkably different? *Neuroscience & Biobehavioral Reviews* 34:104496.

Reader SM, Biro D. 2010 Experimental identification of social learning in wild animals. *Learning & Behavior* 38(3):265–283.

Redshaw J, Taylor AH, Suddendorf T. 2017. Flexible planning in ravens? *Trends in Cognitive Sciences* 21(11):821–822.

Reid AK, Staddon JE. 1997. A reader for the cognitive map. *Information Sciences* 100(1):217–228.

Reid AK, Staddon JE. 1998. A dynamic route finder for the cognitive map. *Psychological Review* 105(3):585.

Reid RA, Reid AK. 2005. Route finding by rats in an open arena. *Behavioural Processes* 68(1):51–67.

Reindl E, Bandini E, Tennie C. 2018. The zone of latent solutions and its relation to the classics: Vygotsky and Köhler. In Di Paolo LD, Di Vincenzo F, De Petrillo F, eds., *Evolution of primate social cognition*, 231–248. Cham, Switzerland: Springer.

Rendell L, Whitehead H. 2001. Culture in whales and dolphins. *Behavioral and Brain Sciences* 24(2):309–382.

Renfrew C. 1996. The sapient behaviour paradox: How to test for potential? In Mellars P, Gibson K, eds., *Modelling the early human mind*, 11–15. Cambridge, UK: McDonald Institute.

Renfrew C. 2008. Neuroscience, evolution and the sapient paradox: The factuality of value and of the sacred. *Philosophical Transactions of the Royal Society B* 363(1499):2041–2047.

Renoult L, Irish M, Moscovitch M, Rugg MD. 2019. From knowing to remembering: The semantic-episodic distinction. *Trends in Cognitive Sciences* 23(12):1041–1057.

Rescorla M. 2017. The computational theory of mind. In Zalta EN, ed., *The Stanford encyclopedia of philosophy*. Metaphysics Research Lab, Stanford University, spring 2017.

Rescorla RA. 1980. Simultaneous and successive associations in sensory preconditioning. *Journal of Experimental Psychology: Animal Behavior Processes* 6(3):207.

Rescorla RA. 1988. Pavlovian conditioning: It's not what you think it is. *American Psychologist* 43(3):151.

Rescorla RA. 1990. Instrumental responses become associated with reinforcers that differ in one feature. *Animal Learning & Behavior* 18(2):206–211.

Rescorla RA. 1992. Depression of an instrumental response by a single devaluation of its outcome. *Quarterly Journal of Experimental Psychology B* 44(2):123–136.

Rescorla RA. 1994. A note on depression of instrumental responding after one trial of outcome devaluation. *Quarterly Journal of Experimental Psychology B* 47(1):27–37.

Rescorla RA, Wagner AR. 1972. A theory of Pavlovian conditioning: Variations in the effectiveness of reinforcement and nonreinforcement. In Black AH, Prokasy WF, eds., *Classical conditioning: Current research and theory*. New York: Appleton-Century-Crofts.

Revusky S. 1984. Associative predispositions. In Marler P, Terrace HS, eds., *The biology of learning*, 447–460. Berlin: Springer.

Reynolds B. 1945. A repetition of the Blodgett experiment on "latent learning." *Journal of Experimental Psychology* 35:504–516.

Reznikova ZI. 2007. *Animal intelligence: From individual to social cognition*. Cambridge, UK: Cambridge University Press.

Richards C, Mottley K, Pearce J, Heyes C. 2009. Imitative pecking by budgerigars, *Melopsittacus undulatus*, over a 24 h delay. *Animal Behaviour* 77(5):1111–1118.

Richerson PJ, Boyd R. 2005. *Not by genes alone: How culture transformed human evolution.* Chicago: University of Chicago Press.

Richerson PJ, Boyd R, Henrich J. 2010. Gene-culture coevolution in the age of genomics. *Proceedings of the National Academy of Sciences* 107(2):8985–8992.

Richerson PJ, Boyd R, Paciotti B. 2002. An evolutionary theory of commons management. In Ostrom E, Dietz T, Dolsak N, Stern PC, Stonich S, Weber EU, eds., *The drama of the commons*, 403–442. Washington, DC: National Academic Press.

Rieznik A, Lebedev M, Sigman M. 2017. Dazzled by the mystery of mentalism: The cognitive neuroscience of mental athletes. *Frontiers in Human Neuroscience* 11:287.

Ringbom H. 2007. *Cross-linguistic similarity in foreign language learning.* Berlin: De Gruyter.

Ritchie S. 2015. *Intelligence: All that matters.* London: Hodder & Stoughton.

Rizzolatti G, Craighero L. 2004. The mirror-neuron system. *Annual Review of Neuroscience* 27(1):169–192.

Robbins MM, Ando C, Fawcett KA, Grueter CC, Hedwig D, Iwata Y, Lodwick JL, Masi S, Salmi R, Stoinski TS, et al. 2016. Behavioral variation in gorillas: Evidence of potential cultural traits. *PLoS One* 11(9):e0160483.

Roberts WA, Grant DS. 1976. Studies of short-term memory in the pigeon using the delayed matching to sample procedure. In Medin DL, Roberts WA, Davis RT, eds., *Processes of animal memory*, 79–112. Hillsdale, NJ: Lawrence Erlbaum.

Robinson A. 1999. *The story of writing.* London: Thames & Hudson.

Robinson JS. 1955. The sameness-difference discrimination problem in chimpanzee. *Journal of Comparative and Physiological Psychology* 48(3):195–197.

Robinson JT. 1954. The genera and species of the Australopithecinae. *American Journal of Physical Anthropology* 12(2):181–200.

Rogers AR. 1988. Does biology constrain culture? *American Anthropologist* 90(4):819–831.

Rogoff B. 1990. *Apprenticeship in thinking: Cognitive development in social context.* New York: Oxford University Press.

Roitblat HL, Bever TG, Terrace HS, eds. 1984. *Animal Cognition.* Hillsdale, NJ: Lawrence Erlbaum.

Roper TJ. 1983. Learning as a biological phenomenon. In Halliday TR, Slater PJ, eds., *Animal behaviour: Genes, development and learning*, vol. 3, 178–212. New York: W. H. Freeman.

Roper T. 1984. Response of thirsty rats to absence of water: Frustration, disinhibition or compensation? *Animal Behaviour* 32(4):1225–1235.

Rothstein B, Stolle D. 2008. Political institutions and generalized trust. In Castiglione D, van Deth JW, Wolleb G, eds., *The handbook of social capital*, 273–302. New York: Oxford University Press.

Rozeboom WW. 1957. Secondary extinction of lever-pressing behavior in the albino rat. *Journal of Experimental Psychology* 54(4):280–287.

Rozin P. 1976. The selection of foods by rats, humans, and other animals. In Rosenblatt JS, Hinde RA, Shaw E, Beer C, eds., *Advances in the Study of Behavior*, vol. 6, 21–76. New York: Academic Press.

Rutz C, Sugasawa S, van der Wal JE, Klump BC, St Clair JJ. 2016. Tool bending in New Caledonian crows. *Royal Society Open Science* 3(8):160439.

Saffran JR, Estes KG. 2006. Mapping sound to meaning: Connections between learning about sounds and learning about words. *Advances in Child Development and Behavior* 34:1–38.

Sala G, Gobet F. 2017. Does far transfer exist? Negative evidence from chess, music, and working memory training. *Current Directions in Psychological Science* 26(6):515–520.

Salazar JM. 1968. Gregariousness in young rats. *Psychonomic Science* 10(11):391–392.

Salomon G, Perkins DN. 1987. Transfer of cognitive skills from programming: When and how? *Journal of Educational Computing Research* 3(2):149–169.

Sanz CM, Call J, Boesch C. 2013. *Tool use in animals: Cognition and ecology*. Cambridge, UK: Cambridge University Press.

Sanz CM, Call J, Morgan D. 2009. Design complexity in termite-fishing tools of chimpanzees (*Pan troglodytes*). *Biology Letters* 5:293–296.

Sarnecka BW, Gelman SA. 2004. Six does not just mean a lot: Preschoolers see number words as specific. *Cognition* 92(3):329–352.

Sarnecka BW, Kamenskaya VG, Yamana Y, Ogura T, Yudovina YB. 2007. From grammatical number to exact numbers: Early meanings of "one," "two," and "three" in English, Russian, and Japanese. *Cognitive Psychology* 55(2):136–168.

Scarr-Salapatek S. 1976. An evolutionary perspective on infant intelligence: Species patterns and individual variations. In Lewis M, ed., *Origins of intelligence*, 165–197. New York: Springer.

Schacter DL, Addis DR. 2007. The cognitive neuroscience of constructive memory: Remembering the past and imagining the future. *Philosophical Transactions of the Royal Society B: Biological Sciences* 362(1481):773–786.

Schacter DL, Guerin SA, Jacques PLS. 2011. Memory distortion: An adaptive perspective. *Trends in Cognitive Sciences* 15(10):467–474.

Schaller GB. 1968. Hunting behaviour of the cheetah in the Serengeti National Park, Tanzania. *African Journal of Ecology* 6(1):95–100.

Schlebusch CM, Malmström H, Günther T, Sjödin P, Coutinho A, Edlund H, Munters AR, Vicente M, Steyn M, Soodyall H, et al. 2017. Southern African ancient genomes estimate modern human divergence to 350,000 to 260,000 years ago. *Science* 358(6363):652–655.

Schneiderman N. 1966. Interstimulus interval function of the nictitating membrane response of the rabbit under delay versus trace conditioning. *Journal of Comparative and Physiological Psychology* 62:397–402.

Seed AM, Boogert NJ. 2013. Animal cognition: An end to insight? *Current Biology* 23(2):R67–R69.

Seki Y, Suzuki K, Osawa AM, Okanoya K. 2013. Songbirds and humans apply different strategies in a sound sequence discrimination task. *Frontiers in Psychology* 4:447.

Seligman M. 1970. On the generality of the laws of learning. *Psychological Review* 77(5):406–418.

Selosse MA, Rousset F. 2011. The plant-fungal marketplace. *Science* 333(6044):828–829.

Seward JP. 1949. An experimental analysis of latent learning. *Journal of Experimental Psychology* 39(2):177–186.

Shah A. 2012. Psychological and neuroscientific connections with reinforcement learning. In Wiering M, van Otterlo M, eds., *Reinforcement learning*, 507–537. Berlin: Springer.

Shannon CE. 1950. Programming a computer for playing chess. *London, Edinburgh, and Dublin Philosophical Magazine and Journal of Science* 41(314):256–275.

Shennan S. 2015. Demography and cultural evolution. In *Emerging trends in the social and behavioral sciences: An interdisciplinary, searchable, and linkable resource*, 1–14. New York: Wiley.

Sherry DF. 1984. Food storage by black-capped chickadees: Memory for the location and contents of caches. *Animal Behaviour* 32(2):451–464.

Sherry DF, Galef BG Jr. 1984. Cultural transmission without imitation: Milk bottle opening by birds. *Animal Behavior* 32(3):937–938.

Sherry DF, Schacter DL. 1987. The evolution of multiple memory systems. *Psychological Review* 94(4):439.

Shettleworth SJ. 1972. Constraints on learning. In Lehrman DS, Hinde RA, Shaw E, eds., *Advances in the study of behavior*, vol. 4, 1–68. New York: Academic Press.

Shettleworth SJ. 1975. Reinforcement and the organisation of behavior in golden hamsters: Hunger, environment and food reinforcement. *Journal of Experimental Psychology: Animal Behavior Processes* 1(1):56–87.

Shettleworth SJ. 2010a. Clever animals and killjoy explanations in comparative psychology. *Trends in Cognitive Sciences* 14(11):477–481.

Shettleworth SJ. 2010b. *Cognition, evolution, and behavior*. New York: Oxford University Press.

Shettleworth SJ. 2012. Do animals have insight, and what is insight anyway? *Canadian Journal of Experimental Psychology* 66(4):217.

Shipley BE, Colwill RM. 1996. Direct effects on instrumental performance of outcome revaluation by drive shifts. *Animal Learning & Behavior* 24(1):57–67.

Shippey T. 2014. *The road to Middle-Earth: How J.R.R. Tolkien created a new mythology*. Boston: Houghton Mifflin.

Siegler RS. 1996. *Emerging minds*. New York: Oxford University Press.

Siegler RS. 2006. Microgenetic analysis of learning. In Kuhn D, Siegler RS, Damon W, Lerner RM, eds., *Handbook of child psychology: Cognition, perception, and language*, vol. 2, 464–510. New York: Wiley & Sons.

Silver D, Hubert T, Schrittwieser J, Antonoglou I, Lai M, Guez A, Lanctot M, Sifre L, Kumaran D, Graepel T, et al. 2018. A general reinforcement learning algorithm that masters chess, shogi, and Go through self-play. *Science* 362(6419):1140–1144.

Simons DJ, Chabris CF. 1999. Gorillas in our midst: Sustained inattentional blindness for dynamic events. *Perception* 28(9):1059–1074.

Simoons FJ. 1970. Primary adult lactose intolerance and the milking habit: A problem in biologic and cultural interrelations. *American Journal of Digestive Diseases* 15(8):695–710.

Simoons FJ. 1971. The antiquity of dairying in Asia and Africa. *Geographical Review* 61(3):431–439.

Singer P, Feldon J, Yee BK. 2009. The glycine transporter 1 inhibitor ssr504734 enhances working memory performance in a continuous delayed alternation task in c57bl/6 mice. *Psychopharmacology* 202(1-3):371–384.

Singley MK, Anderson JR. 1989. *The transfer of cognitive skill.* Cambridge, MA: Harvard University Press.

Skinner BF. 1934. The extinction of chained reflexes. *Proceedings of the National Academy of Sciences* 20(4):234.

Skinner BF. 1938. *The behavior of organisms: An experimental analysis.* La Jolla, CA: Copley Publishing Group.

Skinner BF. 1950. Are theories of learning necessary? *Psychological Review* 57:193–216.

Skinner BF. 1984. The phylogeny and ontogeny of behavior. *Behavioral and Brain Sciences* 7(4):669–677.

Slackman EA, Hudson JA, Fivush R. 1986. Actions, actors, links, and goals: The structure of children's event representations. In Nelson K, ed., *Event knowledge: Structure and function in development,* 47–69. Hillsdale, NJ: Lawrence Erlbaum.

Slobodchikoff CN. 2002. Cognition and communication in prairie dogs. In *The cognitive animal: Empirical and theoretical perspectives on animal cognition,* 257–264. Cambridge, MA: MIT Press.

Slocombe KE, Seed AM. 2019. Cooperation in children. *Current Biology* 29(11):R470–R473.

Sloutsky VM. 2010. Mechanisms of cognitive development: Domain-general learning or domain-specific constraints? *Cognitive Science* 34(7):1125–1130.

Snyder WD. 2018. Chimpanzee cracking nut with hammer and anvil. https://creativecommons.org/licenses/by-sa/4.0/deed.en.

Sol D. 2009. Revisiting the cognitive buffer hypothesis for the evolution of large brains. *Biology Letters* 5(1):130–133.

Sousa C, Biro D, Matsuzawa T. 2009. Leaf-tool use for drinking water by wild chimpanzees (*Pan troglodytes*): Acquisition patterns and handedness. *Animal Cognition* 12(1):115–125.

Sowa JF. 1987. Semantic networks. In Shapiro SC, ed., *Encyclopedia of artificial intelligence.* New York: Wiley.

Sowa JF. 2014. *Principles of semantic networks: Explorations in the representation of knowledge.* San Francisco: Morgan Kaufmann.

Spelke ES, Kinzler KD. 2007. Core knowledge. *Developmental Science* 10(1):89–96.

Spence KW. 1932. The order of eliminating blinds in maze learning by the rat. *Journal of Comparative Psychology* 14(1):9–27.

Spence KW. 1947. The role of secondary reinforcement in delayed reward learning. *Psychological Review* 54:1–8.

Spence KW. 1950. Cognitive versus stimulus-response theories of learning. *Psychological Review* 57(3):159.

Spence KW, Lippitt R. 1946. An experimental test of the sign-gestalt theory of trial and error learning. *Journal of Experimental Psychology* 36(6):491.

Spence KW, Shipley WC. 1934. The factors determining the difficulty of blind alleys in maze learning by the white rat. *Journal of Comparative Psychology* 17(3):423–436.

Speth J. 2004. News flash: Negative evidence convicts Neanderthals of gross mental incompetence. *World Archaeology* 36(4):519–526.

Staddon JE. 2001. *The new behaviorism: Mind, mechanism and society.* Hove, East Sussex, UK: Psychology Press.

Stamp Dawkins M. 2008. The science of animal suffering. *Ethology* 114(10):937–945.

Stanovich KE. 1986. Matthew effects in reading: Some consequences of individual differences in the acquisition of literacy. *Reading Research Quarterly* 189(1–2):360–407.

Stansbury AL, Janik VM. 2019. Formant modification through vocal production learning in gray seals. *Current Biology* 29(13):2244–2249.

Stearns SC, ed. 1987. *The evolution of sex and its consequences.* New York: Springer.

Steffen W, Sanderson RA, Tyson PD, Jäger J, Matson PA, Moore III B, Oldfield F, Richardson K, Schellnhuber HJ, Turner BL, et al. 2006. *Global change and the Earth system: A planet under pressure.* Berlin: Springer.

Stephens DW, Krebs JR. 1986. *Foraging theory.* Princeton, NJ: Princeton University Press.

Sterelny K. 2003. *Thought in a hostile world: The evolution of human cognition.* Oxford, UK: Wiley-Blackwell.

Sterelny K. 2012. *The evolved apprentice.* Cambridge, MA: MIT Press.

Sterelny K, Joyce R, Calcott B, Fraser B. 2013. *Cooperation and its evolution.* Cambridge, MA: MIT Press.

Sternberg RJ, Grigorenko EL, eds. 2002. *The general factor of intelligence: How general is it?* Hove, East Sussex, UK: Psychology Press.

Steward JH. 1972. *Theory of culture change: The methodology of multilinear evolution.* Champaign: University of Illinois Press.

Stoinski TS, Whiten A. 2003. Social learning by orangutans (*Pongo abelii* and *Pongo pygmaeus*) in a simulated food-processing task. *Journal of Comparative Psychology* 117(3):272.

Stoinski TS, Wrate JL, Ure N, Whiten A. 2001. Imitative learning by captive western lowland gorillas (*Gorilla gorilla gorilla*) in a simulated food-processing task. *Journal of Comparative Psychology* 115(3):272.

Stout D. 2011. Stone toolmaking and the evolution of human culture and cognition. *Philosophical Transactions of the Royal Society B* 366:1050–1059.

Stout D, Toth N, Schick K, Chaminade T. 2008. Neural correlates of Early Stone Age toolmaking: Technology, language and cognition in human evolution. *Philosophical Transactions of the Royal Society B: Biological Sciences* 363(1499):1939–1949.

Straub RO, Seidenberg MS, Bever TG, Terrace HS. 1979. Serial learning in the pigeon. *Journal of the Experimental Analysis of Behavior* 32:137–148.

Straub RO, Terrace HS. 1981. Generalization of serial learning in the pigeon. *Animal Learning & Behavior* 9(4):454–468.

Strimling P, Sjöstrand J, Enquist M, Eriksson K. 2009. Accumulation of independent cultural traits. *Theoretical Population Biology* 76:77–83.

Suddendorf T, Addis DR, Corballis MC. 2009. Mental time travel and the shaping of the human mind. *Philosophical Transactions of the Royal Society B* 364(1521):1317–1324.

Suddendorf T, Corballis MC. 1997. Mental time travel and the evolution of the human mind. *Genetic, Social, and General Psychology Monographs* 123(2):133–167.

Suddendorf T, Corballis MC. 2007. The evolution of foresight: What is mental time travel, and is it unique to humans? *Behavioral and Brain Sciences* 30(03):299–313.

Sugiyama Y, Koman J. 1979. Tool-using and making behavior in wild chimpanzees at Bossou, Guinea. *Primates* 20:513–524.

Sun R. 2001. *Duality of the mind: A bottom-up approach toward cognition.* Hove, East Sussex, UK: Psychology Press.

Sundqvist A, Nordqvist E, Koch FS, Heimann M. 2016. Early declarative memory predicts productive language: A longitudinal study of deferred imitation and communication at 9 and 16 months. *Journal of Experimental Child Psychology* 151:109–119.

Sutton RS, Barto AG. 1981. Toward a modern theory of adaptive networks: Expectation and prediction. *Psychological Review* 88:135–140.

Sutton RS, Barto AG. 2018. *Reinforcement learning*. Cambridge, MA: MIT Press.

Suzuki TN, Wheatcroft D, Griesser M. 2016. Experimental evidence for compositional syntax in bird calls. *Nature Communications* 7(1):1–7.

Szathmáry E. 2015. Toward major evolutionary transitions theory 2.0. *Proceedings of the National Academy of Sciences* 112(33):10104–10111.

Taatgen NA. 2013. The nature and transfer of cognitive skills. *Psychological Review* 120(3):439.

Tamis-LeMonda CS, Bornstein MH, Cyphers L, Toda S, Ogino M. 1992. Language and play at one year: A comparison of toddlers and mothers in the United States and Japan. *International Journal of Behavioral Development* 15(1):19–42.

Tattersall I. 1995. *The fossil trail: How we know what we know about human evolution*. New York: Oxford University Press.

Taylor AH, Hunt G, Medina F, Gray R. 2009. Do New Caledonian crows solve physical problems through causal reasoning? *Proceedings of the Royal Society B* 276(1655):247–254.

Taylor AH, Knaebe B, Gray RD. 2012. An end to insight? New Caledonian crows can spontaneously solve problems without planning their actions. *Proceedings of the Royal Society B* 279(1749):4977–4981.

Tebbich S, Sterelny K, Teschke I. 2010. The tale of the finch: Adaptive radiation and behavioural flexibility. *Philosophical Transactions of the Royal Society B: Biological Sciences* 365(1543):1099–1109.

Tebbich S, Taborsky M, Fessl B, Blomqvist D. 2001. Do woodpecker finches acquire tool-use by social learning? *Proceedings of the Royal Society B: Biological Sciences* 268(1482):2189–2193.

ten Cate C. 1984. The influence of social relations on the development of species recognition in zebra finches. *Behaviour* 91:263–285.

Tennie C, Call J, Tomasello M. 2006. Push or pull: Imitation vs. emulation in great apes and human children. *Ethology* 112:1159–1169.

Tennie C, Call J, Tomasello M. 2010. Evidence for emulation in chimpanzees in social settings using the floating peanut task. *PLoS One* 5(5):e10544.

Terkel J. 1995. Cultural transmission in the black rat: Pine cone feeding. *Advances in the Study of Behavior* 24:119–154.

Terkel J. 1996. Cultural transmission of feeding behavior in the black rat (*Rattus rattus*). In Heyes C, Galef BG Jr., eds., *Social learning in animals: The roots of culture*. New York: Academic Press.

Terrace HS. 1986. A nonverbal organism's knowledge of ordinal position in a serial learning task. *Journal of Experimental Psychology: Animal Behavior Processes* 12(3):203–214.

Terrace HS. 2005. The simultaneous chain: A new approach to serial learning. *Trends in Cognitive Sciences* 9(4):202–210.

Terrace HS, Metcalfe J, eds. 2005. *The missing link in cognition: Origins of self-reflective consciousness*. New York: Oxford University Press.

Terrace HS, Petitto LA, Sanders RJ, Bever TG. 1979. Can an ape create a sentence? *Science* 206(4421):891–902.

Thelen E, Smith LB. 2006. Dynamic systems theories. In Lerner RM, ed., *Handbook of child psychology: Theoretical models of human development*, vol. 1, 258–312. 6th ed. New York: Wiley.

Thistlethwaite D. 1951. A critical review of latent learning and related experiments. *Psychological Bulletin* 48(2):97–129.

Thorndike EL. 1911. *Animal intelligence: Experimental studies*. New York: Macmillan.

Thornton A, McAuliffe K. 2006. Teaching in wild meerkats. *Science* 313(5784):227–229.

Thorpe W. 1963. *Learning and instinct in animals*. Cambridge, MA: Harvard University Press.

Thouless C, Fanshawe J, Bertram B. 1989. Egyptian vultures *Neophron percnopterus* and ostrich *Struthio camelus* eggs: The origins of stone-throwing behaviour. *Ibis* 131(1):9–15.

Thrailkill EA, Bouton ME. 2016. Extinction of chained instrumental behaviors: Effects of consumption extinction on procurement responding. *Learning & Behavior* 44(1):85–96.

Timberlake W. 1983. The functional organization of appetitive behavior: Behavior systems and learning. In Zeiler D, Harzem P, eds., *Advances in the analysis of behavior: Biological factors in learning*, vol. 3, 177–221. New York: Wiley.

Timberlake W. 1994. Behavior systems, associationism, and Pavlovian conditioning. *Psychonomic Bulletin & Review* 1(4):405–420.

Timberlake W. 2001. Motivational modes in behavior systems. In Mowrer RR, Klein SB, eds., *Handbook of contemporary learning theories*, 155–209. Hillsdale, NJ: Lawrence Erlbaum.

Timberlake W, Silva KM. 1995. Appetitive behavior in ethology, psychology, and behavior systems. In Thompson NS, ed., *Perspectives in ethology*, vol. 11, 211–253. New York: Plenum.

Timberlake W, Wahl G, King DA. 1982. Stimulus and response contingencies in the misbehavior of rats. *Journal of Experimental Psychology: Animal Behavior Processes* 8(1):62–85.

Tinbergen N. 1951. *The study of instinct*. London: Oxford University Press.

Tinbergen N. 1963. On aims and methods of ethology. *Zeitschrift für Tierpsychologie* 20:410–433.

Tolman EC. 1933. Sign-gestalt or conditioned reflex. *Psychological Review* 40(3):246–255.

Tolman EC. 1948. The cognitive map in rats and man. *Psychological Review* 55:189–208.

Tolman EC, Gleitman H. 1949. Studies in learning and motivation: I. Equal reinforcements in both end-boxes, followed by shock in one end-box. *Journal of Experimental Psychology* 39(6):810–819.

Tolman EC, Honzik CH. 1930a. "Insight" in rats. *University of California Publications in Psychology* 4:215–232.

Tolman EC, Honzik CH. 1930b. Introduction and removal of reward, and maze performance in rats. *University of California Publications in Psychology* 4:257–275.

Tomasello M. 1996. Do apes ape? In Heyes C, Galef BG Jr., eds., *Social learning in animals: The roots of culture*, 319–346. New York: Academic Press.

Tomasello M. 1999. *The cultural origins of human cognition*. Cambridge, MA: Harvard University Press.

Tomasello M. 2003. *Constructing a language: A usage-based theory of language acquisition*. Cambridge, MA: Harvard University Press.

Tomasello M. 2007. Acquiring linguistic constructions. In Siegler R, Kuhn D, eds., *Handbook of child psychology: Cognitive development*, vol. 2, 255–298. Hoboken, NJ: Wiley Online Library.

Tomasello M. 2009. *Why we cooperate*. Cambridge, MA: MIT Press.

Tomasello M. 2010. *Origins of human communication*. Cambridge, MA: MIT Press.

Tomasello M. 2014. *A natural history of human thinking*. Cambridge, MA: Harvard University Press.

Tomasello M, Call J. 1997. *Primate cognition*. New York: Oxford University Press.

Tomasello M, Call J. 2004. The role of humans in the cognitive development of apes revisited. *Animal Cognition* 7(4):213–215.

Tomasello M, Carpenter M, Call J, Behne T, Moll H. 2005. Understanding and sharing intentions: The origins of cultural cognition. *Behavioral and Brain Sciences* 28(5):675–691.

Tomasello M, Davis Dasilva M, Camak L. 1987. Observational learning of tool-use by young chimpanzees. *Human Evolution* 2(2):175–183.

Tomasello M, Kruger AC, Ratner HH. 1993. Cultural learning. *Behavioral and Brain Sciences* 16:495–511.

Tomasello M, Rakoczy H. 2003. What makes human cognition unique? From individual to shared to collective intentionality. *Mind & Language* 18(2):121–147.

Tomasello M, Savage-Rumbaugh S, Kruger AC. 1993. Imitative learning of actions on objects by children, chimpanzees and encultured chimpanzees. *Child Development* 64:1688–1705.

Tomonaga M, Matsuzawa T. 2000. Sequential responding to Arabic numerals with wild cards by the chimpanzee (*Pan troglodytes*). *Animal Cognition* 3(1):1–11.

Tooby J, Cosmides L. 1989. Evolutionary psychology and the generation of culture, part 1. *Ethology and Sociobiology* 10:29–49.

Tooby J, Cosmides L. 1992. The psychological foundations of culture. In Barkow JH, Cosmides L, Tooby J, eds., *The adapted mind: Evolutionary psychology and the generation of culture*, 19–136. New York: Oxford University Press.

Topál J, Byrne RW, Miklósi A, Csányi V. 2006. Reproducing human actions and action sequences: "Do as I do!" in a dog. *Animal Cognition* 9(4):355–367.

Toth N, Schick KD, Savage-Rumbaugh ES, Sevcik RA, Rumbaugh DM. 1993. Pan the tool-maker: Investigations into the stone tool-making and tool-using capabilities of a bonobo (*Pan paniscus*). *Journal of Archaeological Science* 20(1):81–91.

Tulving E. 1972. Episodic and semantic memory. In Tulving E, Donaldson W, eds., *Organization of memory*, 381–402. New York: Academic Press.

Tulving E. 2001. Episodic memory and common sense: How far apart? *Philosophical Transactions of the Royal Society. Series B* 356(1413):1505–1515.

Tulving E. 2002. Episodic memory: From mind to brain. *Annual Review of Psychology* 53(1): 1–25.

Tulving E. 2005. Episodic memory and autonoesis: Uniquely human? In Terrace HS, Metcalfe J, eds., *The missing link in cognition: Origins of self-reflective consciousness*, 3–56. New York: Oxford University Press.

Vale GL, McGuigan N, Burdett E, Lambeth SP, Lucas A, Rawlings B, Schapiro SJ, Watson SK, Whiten A. 2020. Why do chimpanzees have diverse behavioral repertoires yet lack more complex cultures? Invention and social information use in a cumulative task. *Evolution and Human Behavior* 42(3):247–258.

Vallortigara G. 2004. Visual cognition and representation in birds and primates. In Rogers LJ, Kaplan G, eds., *Comparative vertebrate cognition*, 57–94. Berlin: Springer.

van de Waal E, Borgeaud C, Whiten A. 2013. Potent social learning and conformity shape a wild primate's foraging decisions. *Science* 340(6131):483–485.

Van Haaren F, De Bruin JP, Heinsbroek RP, Van de Poll NE. 1985. Delayed spatial response alternation: Effects of delay-interval duration and lesions of the medial prefrontal cortex on response accuracy of male and female wistar rats. *Behavioural Brain Research* 18(1):41–49.

VanLehn K. 1996. Cognitive skill acquisition. *Annual Review of Psychology* 47(1):513–539.

van Schaik CP. 2010. Social learning and culture in animals. In Kappeler P, ed., *Animal behaviour: Evolution and mechanisms*, 623–653. Springer.

van Schaik CP, Ancrenaz M, Borgen G, Galdikas B, Knott CD, Singleton I, Suzuki A, Utami SS, Merrill M. 2003. Orangutan cultures and the evolution of material culture. *Science* 299(5603):102–105.

Vasey DE. 2002. *An ecological history of agriculture 10,000 BC–AD 10,000*. West Lafayette, IN: Purdue University Press.

Vauclair J. 1982. Sensorimotor intelligence in human and non-human primates. *Journal of Human Evolution* 11(3):257–264.

Vaughan W, Greene SL. 1984. Pigeon visual memory capacity. *Journal of Experimental Psychology: Animal Behavior Processes* 10(2):256–271.

Visalberghi E. 1987. Acquisition of nut-cracking behaviour by 2 capuchin monkeys (*Cebus apella*). *Folia Primatologica* 49(3–4):168–181.

Visalberghi E. 1990. Tool use in *Cebus*. *Folia Primatologica* 54(3–4):146–154.

Visalberghi E, Limongelli L. 1994. Lack of comprehension of cause-effect relations in tool-using capuchin monkeys (*Cebus apella*). *Journal of Comparative Psychology* 108(1):15–22.

Voelkl B, Huber L. 2000. True imitation in marmosets. *Animal Behaviour* 60(2):195–202.

Voicu H, Schmajuk N. 2002. Latent learning, shortcuts and detours: A computational model. *Behavioural Processes* 59(2):67–86.

Von Frisch K. 1967. *The dance language and orientation of bees*. Cambridge, MA: Harvard University Press.

Voorhees B, Read D, Gabora L. 2020. Identity, kinship, and the evolution of cooperation. *Current Anthropology* 61(2):194–218.

Vrba ES. 1995. The fossil record of African antelopes (Mammalia, Bovidae) in relation to human evolution and paleoclimate. In Vrba E, Denton G, Partridge T, Burckle L, eds., *Paleoclimate evolution, with emphasis on human origins*, 385–424. New Haven, CT: Yale University Press.

Vygotsky LS. 1962. Language and thought. Cambridge, MA: MIT Press.

Vygotsky LS. 1978. *Mind in society: The development of higher psychological processes*. Cambridge, MA: Harvard University Press.

Walker CBF. 1987. *Cuneiform: Reading the past*. London: British Museum Press.

Walker MM. 1984. Magnetic sensitivity and its possible physical basis in the yellowfin tuna, *Thunnus albacares*. In McCleave JD, Arnold GP, Dodson JJ, Neill WH, eds., *Mechanisms of Migration in Fishes*, 125–141. Boston: Springer.

Ward CV. 2002. Interpreting the posture and locomotion of *Australopithecus afarensis*: Where do we stand? *American Journal of Physical Anthropology* 119(S35):185–215.

Ward G, Allport A. 1997. Planning and problem solving using the five disc Tower of London task. *Quarterly Journal of Experimental Psychology* 50(1):49–78.

Ward-Robinson J, Hall G. 1996. Backward sensory preconditioning. *Journal of Experimental Psychology: Animal Behavior Processes* 22(4):395.

Ward-Robinson J, Hall G. 1998. Backward sensory preconditioning when reinforcement is delayed. *Quarterly Journal of Experimental Psychology* 51(4):349–362.

Warneken F, Chen F, Tomasello M. 2006. Cooperative activities in young children and chimpanzees. *Child Development* 77(3):640–663.

Warner RR. 1988. Traditionality of mating-site preferences in a coral reef fish. *Nature* 335(6192):719.

Washburn MF. 1908. *The animal mind: A textbook of comparative psychology*. New York: Macmillan.

Watanabe S, Sakamoto J, Wakita M. 1995. Pigeons' discrimination of paintings by Monet and Picasso. *Journal of the Experimental Analysis of Behavior* 63(2):165–174.

Watrin JP, Darwich R. 2012. On behaviorism in the cognitive revolution: Myth and reactions. *Review of General Psychology* 16(3):269–282.

Wegner DM. 1987. Transactive memory: A contemporary analysis of the group mind. In Mullen B, Goethals GR, eds., *Theories of group behavior*, 185–208. New York: Springer.

Weigl PD, Hanson EV. 1980. Observational learning and the feeding behavior of the red squirrel *Tamiasciurus hudsonicus*: The ontogeny of optimization. *Ecology* 61(2):213–218.

Weisman R, Duder C, von Konigslow R. 1985. Representation and retention of three-event sequences in pigeons. *Learning and Motivation* 16(3):239–258.

Weisman R, Wasserman E, Dodd P, Larew MB. 1980. Representation and retention of two-event sequences in pigeons. *Journal of Experimental Psychology: Animal Behavior Processes* 6(4):312.

Werner EE, Hall DJ. 1974. Optimal foraging and the size selection of prey by the bluegill sunfish (*Lepomis macrochirus*). *Ecology* 55(5):1042–1052.

Westergaard GC, Lundquist AL, Haynie MK, Kuhn HE, Suomi SJ. 1998. Why some capuchin monkeys (*Cebus apella*) use probing tools (and others do not). *Journal of Comparative Psychology* 112(2):207–211.

Westneat DF, Fox C, eds. 2010. *Evolutionary behavioral ecology*. New York: Oxford University Press.

White TD, Asfaw B, Beyene Y, Haile-Selassie Y, Lovejoy CO, Suwa G, WoldeGabriel G. 2009. *Ardipithecus ramidus* and the paleobiology of early hominids. *Science* 326(5949):64–86.

Whiten A. 1997. The Machiavellian mindreader. In Whiten A, Byrne R, eds., *Machiavellian intelligence: Extensions and evaluations*, vol. 2, 144–173. Cambridge, UK: Cambridge University Press.

Whiten A. 1998. Imitation of sequential structure of actions by chimpanzees. *Journal of Comparative Psychology* 112:270–281.

Whiten A. 2000. Primate culture and social learning. *Cognitive Science* 24(3):477–508.

Whiten A, Byrne RW. 1988. *The Machiavellian intelligence hypotheses: Editorial*. Oxford, UK: Clarendon Press.

Whiten A, Goodall J, McGrew WC, Nishida T, Reynolds V, Sugiyama Y, Tutin CEG, Wrangham RW, Boesch C. 1999. Cultures in chimpanzees. *Nature* 399:682–685.

Wich SA, Atmoko SSU, Setia TM, van Schaik CP. 2010. *Orangutans: Geographic variation in behavioral ecology and conservation*. New York: Oxford University Press.

Widrow B, Stearns SD. 1985. *Adaptive signal processing*. Englewood Cliffs, NJ: Prentice Hall.

Wiering M. 2005. QV(λ)-learning: A new on-policy reinforcement learning algrithm. In *Proceedings of the 7th European Workshop on Reinforcement Learning*, 17–18.

Wiering M, van Otterlo M. 2012. *Reinforcement learning: State-of-the-art*. Berlin: Springer.

Williams BA. 1994. Conditioned reinforcement: Experimental and theoretical issues. *Behavior Analyst* 2:261–285.

Williams BA, Dunn R. 1991. Preference for conditioned reinforcement. *Journal of the Experimental Analysis of Behavior* 55(1):37–46.

Winsler A, Fernyhough C, Montero I. 2000. *Private speech, executive functioning, and the development of verbal self-regulation*. Cambridge, UK: Cambridge University Press.

Wolcott HF. 1991. Propriospect and the acquisition of culture. *Anthropology & Education Quarterly* 22(3):251–273.

Wolpoff MH, Hawks J, Caspari R. 2000. Multiregional, not multiple origins. *American Journal of Physical Anthropology* 112(1):129–136.

Wood D, Wood H, Middleton D. 1978. An experimental evaluation of four face-to-face teaching strategies. *International Journal of Behavioral Development* 1(2):131–147.

Wood P. 1997. A secondary analysis of claims regarding the reflective judgment interview: International consistency, sequentiality, and intradivisional differences in ill-structured problem solving. In Smart JC, ed., *Higher education: Handbook of theory and research*, vol. 12, 243–312. New York: Agathon.

Woodward AL. 2009. Infants' grasp of others' intentions. *Current Directions in Psychological Science* 18(1):53–57.

Woolley AW, Chabris CF, Pentland A, Hashmi N, Malone TW. 2010. Evidence for a collective intelligence factor in the performance of human groups. *Science* 330(6004):686–688.

Wrangham R. 2009. *Catching fire: How cooking made us human*. New York: Basic Books.

Wrangham R, Carmody R. 2010. Human adaptation to the control of fire. *Evolutionary Anthropology* 19(5):187–199.

Wu X, Zhang C, Goldberg P, Cohen D, Pan Y, Arpin T, Bar-Yosef O. 2012. Early pottery at 20,000 years ago in Xianrendong Cave, China. *Science* 336(6089):1696–1700.

Wyckoff LB Jr. 1952. The role of observing responses in discrimination learning: Part I. *Psychological Review* 59(6):431–442.

Wyckoff LB Jr. 1969. The role of observing responses in discrimination learning: In Hendry DP, ed., *Conditioned reinforcement*. Homewood, IL: Dorsey Press.

Wynn K. 1992. Addition and subtraction by human infants. *Nature* 358(6389):749–750.

Wynn T, Coolidge FL. 2004. The expert Neandertal mind. *Journal of Human Evolution* 46(4):467–487.

Wynne C. 2008. Aping language: A skeptical analysis of the evidence for nonhuman primate language. *Skeptic* 13(4):10–15.

Yang X. 1994. Endogenous vs. exogenous comparative advantage and economies of specialization vs. economies of scale. *Journal of Economics* 60(1):29–54.

Yerkes RM, Yerkes DN. 1928. Concerning memory in the chimpanzee. *Journal of Comparative Psychology* 8(3):237.

Yiğit N, Çolak E, Sözen M, Özkurt Ş, Verimli R. 2001. Observations on the feeding biology and behaviour of the fat dormouse, *Glis glis orientalis* Nehring, 1903 (Mammalia: Rodentia) in captivity. *Zoology in the Middle East* 22(1):17–24.

Yotova V, Lefebvre JF, Moreau C, Gbeha E, Hovhannesyan K, Bourgeois S, Bédarida S, Azevedo L, Amorim A, Sarkisian T, et al. 2011. An X-linked haplotype of Neandertal origin is present among all non-African populations. *Molecular Biology and Evolution* 28(7):1957–1962.

Young HF, Greenberg ER, Paton W, Jane JA. 1967. A reinvestigation of cognitive maps. *Psychonomic Science* 9(11):589–590.

Zador AM. 2019. A critique of pure learning and what artificial neural networks can learn from animal brains. *Nature Communications* 10(1):1–7.

Zakharov VB, Mal'kovskii MG, Mostyaev AI. 2019. On solving the problem of 7-piece chess endgames. *Programming and Computer Software* 45(3):96–98.

Zeldin RK, Olton DS. 1986. Rats acquire spatial learning sets. *Journal of Experimental Psychology: Animal Behavior Processes* 12(4):412.

Zentall TR. 2004. Action imitation in birds. *Animal Learning & Behavior* 32(1):15–23.

Zentall TR. 2006a. Imitation: Definitions, evidence, and mechanisms. *Animal Cognition* 9(4):335–353.

Zentall TR. 2006b. Mental time travel in animals: A challenging question. *Behavioural Processes* 72(2):173–183.

Zentall TR. 2011. Perspectives on observational learning in animals. *Journal of Comparative Psychology* 126(2):114–128.

Zentall TR, Sutton JE, Sherburne LM. 1996. True imitative learning in pigeons. *Psychological Science* 7(6):343–346.

Zilhão J, d'Errico F, Bordes JG, Lenoble A, Texier JP, Rigaud JP. 2006. Analysis of Aurignacian interstratification at the Châtelperronian-type site and implications for the behavioral modernity of Neandertals. *Proceedings of the National Academy of Sciences* 103(33):12643–12648.

Zimmerman DW. 1969. Concurrent schedules of primary and conditioned reinforcement in rats. *Journal of the Experimental Analysis of Behavior* 12(2):261–268.

Zimmerman J. 1963. Technique for sustaining behavior with conditioned reinforcement. *Science* 142(3593):682–684.

Zimmerman J, Hanford PV. 1966. Sustaining behavior with conditioned reinforcement as the only response-produced consequence. *Psychological Reports* 19(2):391–401.

INDEX

A NOTE ON THE TYPE

This book has been composed in Arno, an Old-style serif typeface in the classic Venetian tradition, designed by Robert Slimbach at Adobe.

CPSIA information can be obtained
at www.ICGtesting.com
Printed in the USA
JSHW042251080123
35864JS00001B/1